Electrical Conductivity in Ceramics and Glass

PART A

CERAMICS AND GLASS: SCIENCE AND TECHNOLOGY

A Series of Monographs

Edited by

JOHN B. WACHTMAN, JR.

Inorganic Materials Division
National Bureau of Standards
Washington, D.C.

Volume 1
Radome Engineering Handbook: Design and Principles
edited by J. D. Walton, Jr.

Volume 2
Physics of Electronic Ceramics *(in two parts)*
edited by L. L. Hench and D. B. Dove

Volume 3
Characterization of Ceramics
edited by L. L. Hench and R. W. Gould

Volume 4
Electrical Conductivity of Ceramics
edited by N. M. Tallan

IN PREPARATION
Ceramic Processing
edited by P. J. Jorgensen

Mechanical Properties of Ceramics
edited by J. B. Wachtman, Jr.

Electrical Conductivity in Ceramics and Glass

(*in two parts*)

PART A

Edited by N. M. Tallan

Department of the Air Force
Aerospace Research Laboratories
Wright-Patterson Air Force Base, Ohio

MARCEL DEKKER, INC. New York   1974

REF
QC
610.4
.T34
pt. A

COPYRIGHT © 1974 by MARCEL DEKKER, INC.

ALL RIGHTS RESERVED

Neither this book nor any part may be reproduced or transmitted in any form or by any means, electronic or mechanical, including photocopying, microfilming, and recording, or by any information storage and retrieval system, without permission in writing from the publisher.

MARCEL DEKKER, INC.
305 East 45th Street, New York, New York 10017

LIBRARY OF CONGRESS CATALOG CARD NUMBER: 73-82195
ISBN: 0-8247-6087-5

Current printing (last digit):
10 9 8 7 6 5 4 3 2 1

PRINTED IN THE UNITED STATES OF AMERICA

CERAMICS AND GLASS: SCIENCE AND TECHNOLOGY

Series Editor:

J. B. WACHTMAN, JR.
Chief, Inorganic Materials Division
National Bureau of Standards

Editorial Board

C. L. BABCOCK
University of Arizona, U.S.

R. L. COBLE
Massachusetts Institute
of Technology, U.S.

I. B. CUTLER
University of Utah, U.S.

A. D. FRANKLIN
National Bureau of Standards,
U.S.

W. HALLER
National Bureau of Standards,
U.S.

N. J. KREIDL
University of Missouri
at Rolla, U.S.

M. NAKAHIRA
National Institute for
Researches in Inorganic
Materials, Japan

H. J. OEL
Institut für
Werkstoffwissenschaften III,
Germany

J. A. PASK
University of California
at Berkeley, U.S.

J. P. ROBERTS
University of Leeds, England

R. ROY
Pennsylvania State
University, U.S.

S. SAITO
Tokyo Institute of
Technology, Japan

E. C. SUBBARAO
Indian Institute of
Technology at Kanpur, India

M. TASHIRO
Kyoto University, Japan

CONTENTS

List of Contributors vii
Contents of Part B ix
Foreword xi
Preface xvii

1. ELECTRICAL TRANSPORT: GENERAL CONCEPTS 1
    David Adler
    I.   Introduction 1
    II. Fundamental Solid-State Theory 5
    III. Boltzmann Transport Theory 19
    IV. Kubo Transport Theory 32
         References 33

2. EXPERIMENTAL TECHNIQUES 35
    Robert N. Blumenthal and Martin A. Seitz
    I.   Introduction 36
    II. General Experimental Considerations 38
    III. Thermodynamic Measurements 60
    IV. Electrical Conductivity 93
    V.   Transference Measurements of Mixed Conductors . . . 112
    VI. Thermoelectric Measurements 143
    VII. Mobility 152
         References 168

3. DEFECT STRUCTURE OF CERAMIC MATERIALS 179
    R. J. Brook
    I.   Introduction 180
    II. Types of Defect 181
    III. Origin of Point Defects 185

IV.	Defects and Electron Energy Levels	201
V.	Defect Equilibria in Crystals	213
VI.	Microstructural Effects	242
VII.	Defects in Glasses	248
VIII.	Defects and Electrical Conductivity	250
IX.	Conclusion	264
	Appendix	265
	References	266

4. ELECTRONIC CONDUCTION 269

J. M. Wimmer and I. Bransky

I.	Introduction	270
II.	Theoretical Forms of the Drift Mobility	271
III.	Charge-Carrier Concentration	279
IV.	Experimental Determination of the Charge-Carrier Concentration and Drift Mobility	287
V.	Analysis of Electrical-Conductivity and Thermal emf Measurements for Two Types of Charge Carriers	293
VI.	Examples in Ceramic Materials	296
	References	307

CONTRIBUTORS TO PART A

DAVID ADLER, Department of Electrical Engineering, Massachusetts Institute of Technology, Cambridge, Massachusetts

ROBERT N. BLUMENTHAL, Department of Mechanical Engineering, Marquette University, Milwaukee, Wisconsin

I. BRANSKY, Ministry of Defense, Haifa, Israel

R. J. BROOK, Atomic Energy Research Establishment, Materials Development Division, Harwell, Didcot, Berkshire, U. K.

MARTIN A. SEITZ, Department of Mechanical Engineering, Marquette University, Milwaukee, Wisconsin

J. M. WIMMER, Aerospace Research Laboratories (LL), Metallurgy and Ceramics Research Laboratory, Wright-Patterson Air Force Base, Ohio

CONTENTS OF PART B

CONTROLLED-VALENCE EFFECTS IN ELECTRONIC CONDUCTORS, J. Bruce Wagner, Jr., Department of Materials Science, Northwestern University, Evanston, Illinois

HIGHLY CONDUCTING CERAMICS AND THE CONDUCTOR — INSULATOR TRANSITION, R. W. Vest and J. M. Honig, Purdue University, Department of Electrical Engineering, Lafayette, Indiana

IONIC CONDUCTIVITY AND ELECTROCHEMISTRY OF CRYSTALLINE CERAMICS, John W. Patterson, Department of Metallurgy, Iowa State University, Ames, Iowa

CONDUCTIVITY OF GLASSES AND OTHER AMORPHOUS SOLIDS, John D. Mackenzie, School of Engineering and Applied Science, University of California, Los Angeles, California

MICROSTRUCTURAL AND POLYPHASE EFFECTS, J. M. Wimmer, H. C. Graham, and N. M. Tallan, Aerospace Research Laboratories (LL), Metallurgy and Ceramics Research Laboratory, Wright-Patterson Air Force Base, Ohio

CUMULATIVE AUTHOR INDEX

CUMULATIVE SUBJECT INDEX

FOREWORD

Ceramics and glass are ancient materials which continue to be used in the old ways but which have also been refined and developed by modern science and technology into many new and useful types of material. Ceramics and glass can be manufactured with unique chemical, thermal, electrical, magnetic, optical, mechanical, and mass transport properties. Their applications are much too long to list here in detail, but some examples will serve to illustrate the great diversity and importance of these applications. Special chemical and thermal properties are exploited in furnaces and crucibles which are essential to the refining, alloying, and casting of metals and to other types of high temperature processing. Special electrical properties form the basis for applications as insulators, dielectrics, piezoelectric transducers, and an ever-increasing family of semiconductor devices. Magnetic ceramics include cores for computer memories and high-coercive-force permanent magnets for electric motors. Special optical properties of some ceramics make them suitable for laser hosts, modulators, infrared windows, etc. The high elastic moduli and hardness of certain ceramics make them suitable for special applications requiring mechanical stability such as gyroscope mounts and wear surfaces while the extremely low thermal expansion achievable is appropriate to telescope mirrors and to applications involving thermal shock. The high intrinsic strength of many ceramics has never been fully exploited because of their brittleness and accompanying sensitivity to flaws, but great improvements in the strength of bulk commercial ceramics and the technique of surface compression combine to offer much promise in this area while the recent development of high-strength, high-temperature ceramic fibers opens new possibilities for composite

materials. The high speed of sound in certain ceramics has led to
their successful use in armor and the low damping characteristic of
many ceramics makes them useful as delay lines. The great range of
rates of mass transport achievable in ceramics makes them useful as
protective coatings at the extreme of low diffusion and makes them
promising as fuel cell components at the other extreme of high diffusion rates. The thermal stability of certain nuclear ceramics has
led to their adoption as fuel elements for most power reactors.

Additional examples could be quoted to illustrate proven uses
of ceramics in modern technology and the promise of other uses to
come, but the continuing improvement of traditional ceramics with
large volume use should also be noted. Glass, electrical, and sanitary whitewares, structural clay products, enamels, and chinaware
are still benefiting from science and technology through improved
properties and reliability.

Ceramic materials are of great interest to scientists quite
apart from their many practical applications. Certain phenomena
occurring in ceramics lie at the frontier of science and stimulate
scientific work of the highest sophistication and quality. These
phenomena include the detailed behavior of superconducting polycrystalline oxides which are semiconductors at room temperature. The
voltage and time-dependent behavior of semiconducting glasses is not
well understood, and the question of the structure of glasses and
its relation to their properties remains the basis of an active and
challenging field of science. The discovery of the widespread occurrence of phase separation in glass and the use of controlled nucleation and controlled crystallization in glass to produce crystalline
ceramics has raised new stimulating scientific questions as well as
enriching technology. The range of behavior from glass to crystalline ceramics is so rarely continuous that the term "ceramic" is
today frequently used to include the whole range of these materials.
The chemistry of crystalline ceramics continues to provide scientific
challenges as well as the discovery of useful new materials. Layer
structures, stacking-block structures, magnetic structures, the

extent and character of solid solution, the type and interaction of point defects, and the structure of nonstoichiometric solids all need to be systematized and quantitatively understood. The chemistry of surfaces, including grain boundaries, is receiving increasing study. Mechanical properties have been extensively studied, but the details of mechanical behavior in specific materials offer many opportunities for research and development. The list of scientific challenges could be extended; ceramic materials are the seat of many phenomena worthy of study.

Some of the above applications are appropriate to a specific types of engineering and some of the scientific problems can be viewed simply as part of the solid state physics and chemistry. It is, however, usually advantageous to view both applications of ceramics and relevant aspects of science together from a materials science viewpoint in which the underlying chemistry, the kinetics of processing to produce a material which is often in a metastable state, and the physics of properties are all taken into account. Understanding the properties of ceramics depends upon understanding how microstructure and small changes of composition affect these properties. Experimental knowledge of the factors controlling properties depends critically upon measurement techniques, particularly upon techniques for measuring properties at high temperature and upon characterization techniques. The series "Ceramics and Glass: Science and Technology" is intended to present a survey of current knowledge of processing characterization, and physical properties of ceramics taking into account the underlying chemistry; it is made up of two parts: (1) textbooks, monographs, and handbooks and (2) selected conference proceedings. The first component in the series is made up of volumes each of which gives a complete treatment of its subject area rather than just a report of progress. Sufficient knowledge of the basic physics, chemistry, and mathematics is assumed to treat each subject on the level of a first-year graduate or advanced undergraduate course in materials science. This level is deliberately chosen to make the textbooks, monographs, and handbooks broadly useful to scientists and engineers in industry and in research and development laboratories as well as to students. The second component of the series is made up of conference proceedings selected for their subject, quality, and timeliness. The growth of technical society meetings in size and diversity of subject matter has caused increasing emphasis to be placed on small, separate topical conferences. Such conferences typically cover the selected topic with a series of invited papers plus a limited number of contributed papers. When successful, the proceedings of such a conference give a state-of-the-art presentation useful for several years until the new information can be assimilated in an organized manner into textbooks and handbooks. Good conferences and their proceedings have an important intermediate place between individual research papers and textbooks. The series "Ceramics and Glass: Science and Technology" accordingly includes selected conference proceedings but recognizes and maintains the distinction between these and the textbook, monograph, and handbook series component.

The present volume, "Electrical Conductivity in Ceramics and Glass" by Norman M. Tallan, is a systematic treatment designed to be suitable as a textbook. A thorough grounding in the general theory of electrical transport is provided in the first chapter which treats electronic and ionic conductivity both from the phenomenological viewpoint and from the detailed quantum mechanical viewpoint. The second chapter presents a correspondingly thorough treatment of experimental techniques including such vital but sometimes neglected topics as current leakage, voltage loss or pickup, gas electrode reactions, maintenance of the state of the sample, and sample preparation. Control of such experimental factors is essential to correct determination of the type, concentration, mobility, and transference number of each conducting species. Succeeding chapters then take up the physical chemistry controlling the types of conducting species and the corresponding conduction mechanisms. These considerations are then applied to controlled valency effects, to highly conducting ceramics, to electrochemistry

FOREWORD

of ceramics, to conductivity in glass, and to effects of microstructure and the existence of more than one phase. The present book thus provides the scientific basis for the technology of conduction in electronic ceramics and so complements another volume in the series, "The Physics of Electronic Ceramics" edited by L. L. Hench and D. B. Dove. The latter is the proceedings of a conference which was designed to survey current knowledge of many types of electronic and technical ceramics (semiconductors, dielectrics, ferroelectrics, magnetic ceramics, and optical materials) in terms of relating their technology to underlying science. The depth and thoroughness of the present book make it a suitable textbook and reference book for fundamental electrical phenomena in these types of materials and a useful companion to the above-mentioned conference proceedings book. Electronic and technical ceramics constitute the most rapidly growing component of the ceramic industry. It is hoped that the present book will be broadly useful to workers and teachers in this field.

John B. Wachtman, Jr.
Chief, Inorganic Materials Division
National Bureau of Standards

PREFACE

Observations and applications of the electric and magnetic properties of materials were among man's early accomplishments, and detailed studies of the phenomena since the eighteenth century played a vital role in the development of modern science. These scientific advances in our understanding of the nature and origin of electric charges, fields, and currents were paralleled, and in many instances preceded, by practical discoveries that permitted the generation, storage, study, and application of electrical power. The tremendous technological expansion of the past twenty-five years has been based in large part on this continuing interplay between new advances in our fundamental understanding of the electrical properties of materials and the discovery of new electronic applications of materials in increasingly more complex devices and systems.

Conductivity studies of ceramics have historically been a major part of the ceramic literature, and there is certainly little doubt that they have contributed greatly to the growth of the ceramic industry. In fact, many of the basic advances in the characterization and fabrication of ceramic materials have been primarily motivated by the compositional and microstructural requirements imposed on them by electrical property applications. In reviewing the older experimental conductivity work, however, one is frequently struck by the limited applicability of much of the data. Too often one finds measurements on specimens that were insufficiently characterized with respect to composition or microstructure, or measurements under improper or undefined thermodynamic conditions, so that distinctions between behavior intrinsic and extrinsic to the material cannot now be unequivocally made. Too often the experimental techniques used were not selected with sufficient care to assure that spurious effects were eliminated for the specific material and environmental conditions under study. There are, however, throughout the ceramic literature good examples of conductivity studies that have contributed significantly to both our knowledge of the electrical properties of these materials and our fundamental understanding of the physical and chemical factors responsible for them. This book is intended as a text to help the advanced student in materials science to understand some of the more recent advances in the theory of electrical conductivity of ceramics and glass and to help him in his own experimental studies. In editing this volume I have selected active leaders in each of the theoretical and experimental areas included, and I have asked them to prepare a text suitable for graduate-level instruction.

The scope of this volume has been intentionally limited to electrical conductivity. Many other related topics, including the dielectric, galvanomagnetic, electro-optic, and electro-acoustic properties of ceramic materials would certainly be of comparable value to the interested reader. Hopefully they will be covered in subsequent volumes within this series.

The first chapter presents a theoretical discussion of the general concepts of charge transport in solids in a form that should be of interest to a wide range of students. Those with a limited background in the physics of charge transport will find a lucid introduction to the macroscopic electrical parameters measured, the microscopic charge densities and charge mobilities calculated, the electronic structure of solids, and the essential features of Boltzmann transport theory. The student who is already somewhat familiar with these topics will, I hope, find new insight and understanding in the development given here.

As noted earlier, one of the greatest hurdles facing the inexperienced student in making meaningful conductivity measurements is the avoidance of the many experimental errors that can arise. Techniques which work well for low impedance specimens at moderately

PREFACE   xix

high temperatures do not extend well to very high impedance specimens at low temperatures, and measurements at very high temperatures are fraught with problems of multiple leakage paths. One has to be scrupulously careful of the materials used as contacts and as furnace or sample holder materials, and each of the measurements which together comprise an extensive conductivity study is generally beset with its own difficulties. The second chapter of this text presents an extended discussion of the experimental details involved in many of these measurements. Sample preparation, compositional equilibration, means of determining departures from stoichiometry and charge carrier concentrations, partial conductivities, and electron mobilities are all carefully and thoroughly reviewed.

One of the vital concerns of any electrical conductivity study is the determination of the charge carrier concentration and its relation to the defect structure of the material. In recent years, in fact, conductivity studies have in turn been used to characterize the predominant defects present and their charge states in many of the more important metal oxides. Chapter 3 discusses primarily the nature and formation of point defects in crystalline compounds, the relationships which describe the interactions of the defect concentrations and their temperature and composition dependences, and the structural aspects of their role as ionic and electronic charge carriers.

Chapters 4 through 8 deal specifically with the electrical conductivity of ceramic materials. Chapter 4 treats overall aspects of the electronic conduction mechanism, tying together many of the general conductivity concepts, the defect structure relationships, and the measurement techniques covered in detail in the introductory chapters. The form of the electronic drift mobility for lattice and ionized impurity scattering processes and for the cases of both weak and strong electron-phonon interactions are discussed from the standpoint of attempts that might be made to utilize experimental conductivity data to distinguish between these conduction mechanisms. This chapter also discusses the separation of the electronic conductivity into its concentration and mobility factors, and separation of the measured total electronic conductivity into its electron and electron hole contributions. Examples of the application of these studies to several classes of oxides of particular interest to the ceramist are given.

The concept of controlled valency in ionic compounds and its usefulness in both explaining the electronic conduction of many ceramic materials and in providing a means of controlling their properties have been especially important to the development of our understanding of electronic conduction. Chapter 5 presents a brief review of these concepts and some experimental examples of their great value to the design and interpretation of an electronic conductivity study.

It is frequently surprising how many materials scientists, even those quite familiar with the insulating and semiconducting properties of metal oxides, are unaware that many ceramic compounds have electronic conductivities comparable to those of metals. This is especially surprising in view of the great potential importance of these materials in many practical electronic, electrochemical, and energy conversion applications. Chapter 6 presents a detailed discussion of the origin of this highly conductive behavior in terms of the band structure of those classes of oxides which are particularly representative and instructive, a review of the experimental results that have been obtained for many of these oxides, and a discussion of the transitions observed in a number of these oxides where conductivity changes of several orders of magnitude occur within very small temperature ranges.

Since so many processes of technological importance, such as mechanical creep, sintering, and oxidation, are diffusion rate controlled over at least some range of environmental conditions, the study of ionic conductivity and its relation to diffusion is especially important. Chapter 7 provides a thorough account of ion migration and its description in diffusion and ionic conductivity terms, the electrolytic behavior of oxides, and the electrochemistry of cells using solid electrolytes for the determination of partial conductivities and thermodynamic data for ceramic compounds.

Chapter 8 presents a discussion of the theoretical and experimental aspects of electrical conductivity in glasses and other amorphous materials. The special considerations introduced by the influence of disorder on the effective concentrations and mobilities of ions and electrons in these materials and some of their emerging important applications are covered in a form that should be quite useful to the student.

Finally, the influences of microstructural features, essential elements of most ceramic products, are considered in Chapter 9. The electrical conductivity of heterogeneous materials is described both from the standpoint of the frequency dependence of the conductivity of composite materials and from the standpoint of the mixing rules which can be applied to them. The relationship between these approaches, their physical significance, and some examples of their application to several materials of general interest are given.

The student will probably find that many of these chapters will stand on their own as reasonably self-contained descriptions of specific aspects of the electrical conductivity of ceramic materials. Together, I hope he will find that they provide a useful text, one which will help him to avoid many of the difficulties inherent in conductivity studies on these materials and which will help him to make his own contributions to our understanding of their properties. The reader will find some apparent overlap between several of the chapters, but in each case I am certain that closer examination will show that different aspects or different essential features of the related topics are in fact covered in each instance. Finally, I wish to thank each of the authors of these chapters who gave so generously of their time and who worked so diligently to make this book possible.

Norman M. Tallan

Chapter 1

ELECTRICAL TRANSPORT:
GENERAL CONCEPTS

David Adler

Department of Electrical Engineering
Massachusetts Institute of Technology
Cambridge, Massachusetts

I. INTRODUCTION 1
II. FUNDAMENTAL SOLID-STATE THEORY 5
    A. Classification of Solids 5
    B. Adiabatic Approximation 7
    C. Energy-Band Theory of Solids 9
    D. Electron Statistics 14
    E. Phonons and Large Polarons 15
    F. Correlations, Mott Transitions and Small Polarons . . 17
III. BOLTZMANN TRANSPORT THEORY 19
    A. Electronic States in a Crystal in the
       Presence of an Applied Electric Field 19
    B. The Boltzmann Equation 22
    C. Hopping Conduction 28
    D. Ionic Conduction 29
IV. KUBO TRANSPORT THEORY 32
    REFERENCES 33

I. INTRODUCTION

An operational definition of electrical conductivity is extremely simple. We apply a small electric field $\mathcal{E}$ to a given material and measure the resulting current flow per unit cross-sectional area, $\underline{J}$. The conductivity is then just the ratio of that component of $\underline{J}$ in the direction of $\mathcal{E}$ to the magnitude of $\mathcal{E}$:

$$\sigma = J_{\mathcal{E}}/\mathcal{E} \tag{1}$$

In general, the current need not be in the direction of the field, and we must define a conductivity tensor

$$J_i = \sum_{j=1}^{3} \sigma_{ij}\mathcal{E}_j \tag{2}$$

where the subscripts refer to three spatial directions (i.e., x,y,z) chosen in any convenient manner. If the current density is measured in A/cm^2 and the field in V/cm, the appropriate unit for conductivity is $ohm^{-1} cm^{-1}$. These are hybrid units, being neither mks nor cgs, but nevertheless are the vernacular.

Experimentally, the electric field is applied by connecting two ends of the sample to metallic electrodes and imposing a voltage difference $\Delta\varphi$ between the electrodes. If the sample has a constant cross-sectional area A, a simple measure of the current flowing in the circuit gives the conductivity via

$$\sigma = \frac{I\ell}{A(\Delta\varphi)} \tag{3}$$

where ℓ is the length of the sample.

It is the electrical current that is being measured in this experiment; i.e., a net flow of charge from one end of the sample to the other. Conductivity and current are macroscopic parameters. However, on the microscopic level any particle that carries charge can contribute to the current, and thus to the conductivity, by simply acquiring a nonzero average velocity. (Conversely, a neutral particle cannot contribute to electrical conductivity, no matter what its velocity.) In general, an electric field will accelerate a charged particle and cause it to attain a velocity $\underline{v}$. We can describe the microscopic contributions to conductivity in terms of a parameter that measures the ratio of the component of velocity in the direction of the applied field to the magnitude of the field:

$$\mu \equiv <\underline{v}>_{\mathcal{E}}/\mathcal{E} \tag{4}$$

μ is known as the mobility, and is measured in cm^2/V-sec (the ratio of cm/sec to V/cm). As is true of conductivity, the mobility is generally a tensor:

$$<v>_i = \sum_{j=i}^{3} \mu_{ij}\mathcal{E}_j \tag{5}$$

Clearly, the contribution of a density n of particles of charge q and mobility μ to the conductivity is

$$\sigma = nq\mu \tag{6}$$

where n is measured in cm^{-3} and q in coulombs (ampere-seconds).

All of these concepts are very simple. But we should bear in mind the distinction that experimentally we measure the macroscopic quantity conductivity, whereas theoretically we calculate the microscopic quantities mobility and density. It is also important to note that electrical conductivity, by its very nature, is not an equilibrium process. We must turn on an external electric field to obtain a current. The external field is a force which destroys thermodynamic equilibrium. It causes a resulting flow of charge, i.e., current, which can be measured if a steady state is achieved. In the absence of an applied field, the sample can be in equilibrium, and no current flows. Thus we can define conductivity as the limit of the quantity in Eq. (1) for small fields:

$$\sigma \equiv \lim_{\mathcal{E}\to 0} \frac{J_{\mathcal{E}}}{\mathcal{E}}$$

$$= (\frac{\partial J_{\mathcal{E}}}{\partial \mathcal{E}})_{\mathcal{E}=0} \tag{7}$$

With this definition, conductivity is independent of field. The small-field regime is thus the region in which the linear relation,

$$J_{\mathcal{E}} = \sigma\mathcal{E} \qquad (8)$$

is obeyed. Equation (8) is known as Ohm's law, and it is valid for most materials up to fields of the order of 10^4 V/cm. At sufficiently high fields, the higher order terms in the Taylor expansion

$$J_{\mathcal{E}} = \sigma\mathcal{E} + \frac{1}{2}\left(\frac{\partial^2 J_{\mathcal{E}}}{\partial \mathcal{E}^2}\right)_{\mathcal{E}=0} \mathcal{E}^2 = + \cdots \qquad (9)$$

become significant and Ohm's law breaks down.

Although conductivity as defined by Eq. (7) has a meaning in the limit of no external field, it can be measured only if a field is applied. Thus it is intrinsically a nonequilibrium concept and requires the formalism of irreversible thermodynamics.

We shall assume that the material is in thermodynamic equilibrium in the absence of an external field. This implies that no additional forces, such as temperature gradients or magnetic fields, exist. It can easily be seen that the presence of such an additional force can lead to an incorrect determination of conductivity; e.g., application of a temperature gradient causes an extra contribution to the current flow. The simplest example of such a coupled-flow phenomenon is the existence of a gradient in the chemical potential of a charged particle. The diffusion that results from the gradient necessarily leads to a transport of charge and thus a current, even in the absence of an applied field. This does not, however, imply infinite conductivity.

In order to handle a problem in irreversible thermodynamics, the concept of local equilibrium must be introduced. It is not a very difficult concept. In order to achieve complete thermodynamic equilibrium, no external forces can be present. The temperature of all substances in thermal contact must become equal, and this implies that no temperature gradients exist. Similarly, no pressure gradients, chemical potential gradients, or electric fields (electrostatic potential gradients) can be present. If we impose a temperature gradient, e.g., by means of keeping one end of a rod at 100°C and the other end at 0°C, we have destroyed equilibrium. However, we can imagine dividing the rod into pieces sufficiently small that after a reasonably long time we can define a steady-state temperature for each piece. This gives us a temperature distribution T(x). Similarly, we can define chemical potential distributions or electrostatic potential distributions for any material. Within each small cell, we assume that the temperature, pressure, and potentials are constant, and thus the appropriate thermodynamic equations of state can be used to evaluate unknown parameters, such as entropy. This gives us a way of defining the entropy of the complete nonequilibrium system by summing the entropies of the small cells. If the external forces are sufficiently small that cells can be defined within which the concepts of temperature, potential, etc., retain a meaning despite the presence of heat and current flows, then we say that local equilibrium exists.

Once we assume local equilibrium, the standard development of irreversible thermodynamics [1] follows, and conductivity theory can be put on a solid ground. The major nonintuitive result emerging from this formulation is the Onsager relation

$$\sigma_{ij} = \sigma_{ji} \qquad (10)$$

i.e., the conductivity tensor is symmetric. Equation (10) is a consequence of every microscopic theory of electrical conduction, but irreversible thermodynamics shows it must be true in general.

II. FUNDAMENTAL SOLID-STATE THEORY

A. CLASSIFICATION OF SOLIDS

A solid is a collection of a large number ($\sim 10^{23}$) of atoms. So, in fact, is a liquid or a gas. Solids can be differentiated from gases by noting that the atoms or molecules in a gas interact only weakly, whereas the atoms in a solid mutually interact to form a state in which all are bound together. This binding results in a cohesive energy, which causes the solid state to have a much lower internal energy than the gaseous state. Cohesive energies of most solids fall in the range of 1-10 eV per molecule. But this is the same range of energies necessary to remove an outer electron from most atoms, so it is possible that some atoms are effectively ionized in the solid. In fact, it is this redistribution of electrons that is responsible for the cohesive energy of solids.

When the outermost electron of an atom or molecule is very tightly bound, the neutral atoms or molecules attract each other only weakly (via the van der Waals attraction, which results from the Coulomb interaction between electric dipoles), and the cohesive energy is very small ($\sim$ 0.1 eV per molecule). These solids consequently melt at low temperatures, and are gases at room temperature. Examples are Ne and Cl_2. We shall not be interested in such "van der Waals" solids. However we should note that they consist of weakly bound neutral particles, and thus have near-zero electrical conductivity.

Molecules such as Cl_2 are held together by a so-called covalent bond, in which an electron from each atom primarily stays in the region between the two atoms and reduces the total energy. This reduction in energy, of the order of 3-7 eV, is primarily quantum mechanical in origin [2]. If an atom has more than one electron that can participate in bonding, it can form covalent bonds with more than one other atom, and an open-ended structure can result. Such a structure is called a covalent solid. Typical examples are Ge, Te, and GaAs.

A smaller reduction in energy of two atoms results when only a single electron is shared between them. If an atom has only one valence electron (e.g., Na) a solid can still result from this weaker bond, essentially because the atom is very easily ionized. Such a bond is called metallic, and cohesive energies of the solids are in the range 1-5 eV per molecule.

ELECTRICAL TRANSPORT: GENERAL CONCEPTS

Finally, in materials made up of two different types of atoms, one with a small number of valence electrons, the other with a nearly complete electronic shell, the lowest energy state will be one in which electrons are transferred from the former type of atom to the latter. For example, in NaCl the Coulomb attraction of an Na^+ ion and a Cl^- ion at small distances greatly outweighs the difference between the energy necessary to remove the outer electron from an Na atom and the energy gained by adding an electron to a Cl atom. A large number of NaCl molecules can form a very low-energy solid by forming an alternating arrangement of Na^+ and Cl^- ions throughout space. NaCl is an example of an ionic solid. The cohesive energies are generally 5-10 eV per molecule.

It is not always clear exactly what is responsible for bonding, and mixed solids are the rule rather than the exception. But electrical conductivity is extremely sensitive to the type of bonding, and it is thus desirable to classify solids as best we can before trying to analyze their electrical properties.

B. ADIABATIC APPROXIMATION

We have seen that a major difference between a solid and a gas is that in the former, electrons can be effectively separated from their atoms, leaving positive ion cores behind. Thus a solid, as opposed to a gas, is composed of positively and negatively charged particles. It is the Coulomb forces between any two such particles that is responsible for the cohesive energy of the solid. A major simplification in analyzing the physical properties of solids results from the fact that, although the electric charge of electrons and ion cores are comparable, their masses differ by a factor of the order of 10^5. Because electrons are so light compared to ion cores, they respond much more quickly to perturbations of equivalent energy. For example, consider the velocity attained by these particles due to thermal vibrations. At room temperature their thermal energy kT is approximately 0.025 eV. If this is the order of the kinetic energy of thermal motion, the velocity of an electron is given by

$$\tfrac{1}{2}m v^2 = 0.025 \text{ eV}$$

where m is the electronic mass. The thermal electronic velocity is thus

$$v \simeq 10^7 \text{ cm/sec} \tag{11}$$

On the other hand, the velocity attained by the ion core is determined from

$$\tfrac{1}{2}MV^2 = 0.025 \text{ eV}$$

where M is the ionic mass. The thermal ionic velocity is thus only

$$V \sim 10^4 \text{ cm/sec} \tag{12}$$

From Eqs. (11) and (12) it is clear that the time it takes an electron to move ~1 Å at thermal velocities is ~10^{-15} sec, whereas it takes an ion core ~10^{-12} sec to cover the same distance. Because of this, it becomes possible to separate electronic from ionic motion in any solid.

The adiabatic approximation [3] assumes that ion cores are too slow to react to the motion of electrons, but that, on the other hand, electrons very rapidly adjust to all ionic motions. This is an important approximation, since it allows us to solve for the electronic energy of a solid when the ion cores are in their equilibrium positions. This simplified problem becomes possible to handle numerically because of the empirical fact that the lowest-energy state of almost all materials appears to be a crystal, i.e., a periodic array of ion cores in which each equivalent ion has precisely the same environment. Consequently, the original 10^{23}-body problem can be reduced to a problem of a small number of atoms (one primitive cell), provided the proper boundary conditions are imposed to guarantee equivalent environments. The ultimate result of these calculations is the energy-band theory of solids, which we shall discuss in Section II,C.

ELECTRICAL TRANSPORT: GENERAL CONCEPTS 9

The problem of the motion of the ion cores is handled empirically by experimentally determining the normal modes of vibration, assuming that the ions undergo only small displacements from their equilibrium positions. The adiabatic approximation then implies that the total energy of the solid is simply the sum of the electronic energy and the vibrational energy of the oscillating ion cores, with negligible coupling between the two motions.

C. ENERGY-BAND THEORY OF SOLIDS

If electrons did not mutually interact, the electronic problem would be essentially trivial. However, electrons repel each other with a Coulomb repulsion of the same magnitude or larger than all other contributions to the energy of the solid. In order to solve the full problem, the one-electron (Hartree-Fock) approximation is usually applied [4]. It is assumed that each electron, as it propagates through the solid, interacts only with the time-averaged negative charge due to all other electrons. This clearly leads to an overestimate of the energy, since it neglects the possibility that other electrons can correlate their motion to avoid being in the same region of space at the same time as the propagating electron. Nevertheless, it can be shown that the one-electron approximation is reasonably accurate for most solids.

In addition to the assumptions of the validity of the adiabatic approximation and the unimportance of correlations, the electronic problem is solved only by postulating that each electron is subjected only to forces that have the periodicity of the crystal lattice. This neglects the defects and impurities that are always present. However, the total density of such static imperfections is small, and corrections can be calculated by perturbation theory after the solution of the periodic problem. The same can be said for dynamic imperfections, such as deviations from periodicity caused by lattice vibrations (phonons). On the other hand, every crystal necessarily has surfaces, which by their presence also destroy perfect periodicity. Since the vast majority of atoms are far from the surface, we might also expect only minor effects due to surfaces, but this is difficult to prove theoretically. Instead, an empirical justification is used. Experimentally, the conductivity of a solid is found to be independent of the size and shape of the crystal. Thus conductivity must not be a sensitive function of the exact boundary conditions, and the real ones can be replaced by fictitious but mathematically convenient ones, namely those which restore perfect periodicity [5].

The major results of the one-electron calculations are as follows: (A) The states available to any electron as it moves through the crystal have energies which lie only in certain ranges, known as bands, and which are separated by regions of forbidden energies, known as gaps; (B) each electron is delocalized, i.e., has equal probability of being found in the vicinity of any equivalent ion core in the crystal; (C) each electronic state in a given band can be labeled by a wave vector $\underline{k}$ which is restricted to lie in a given range of values called a Brillouin zone, whose shape depends only on the lattice structure [6]; (D) the total number of states in an energy band is 2N, where N is the total density of equivalent atoms; and (E) the velocity of an electron in a state $\underline{k}$ is

$$\underline{v} = \frac{1}{\hbar} \nabla_{\underline{k}} E(\underline{k}) \tag{13}$$

where $E(\underline{k})$ is the energy of state $\underline{k}$. These states, labeled by $\underline{k}$, are known as Bloch states. Some of these results are illustrated in Fig. 1.

In addition, several symmetry relations can easily be derived [7]. In particular, for any crystal,

$$E(\underline{k}) = E(-\underline{k}) \tag{14}$$

and

$$[\nabla_{\underline{k}} E(\underline{k})]_{B.Z.} = 0 \tag{15}$$

The latter relation means that the normal component of the gradient

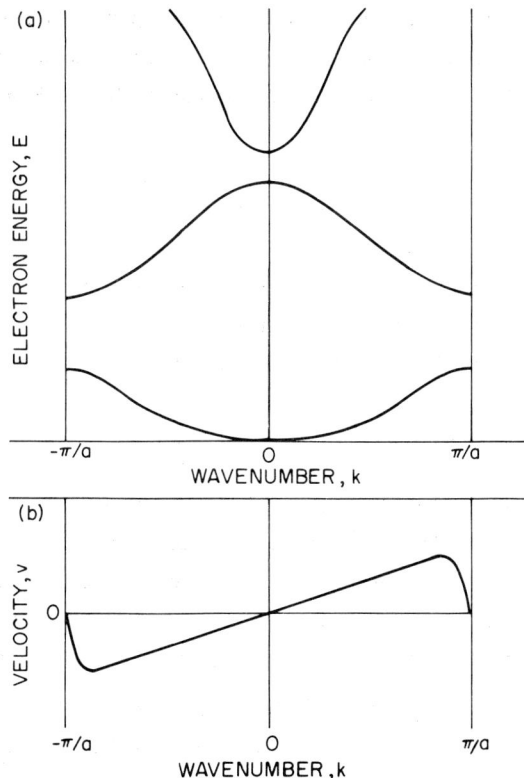

FIG. 1. (a) Sketch of the electronic energy-band structure of a one-dimensional periodic arrangement of atoms as a function of the wavenumber k. Each state is delocalized, in the sense that the electron occupying it has equal probability of being found near any equivalent atom in the crystal. (b) The velocity of electrons in the states of the lowest band in (a).

of $E(\underline{k})$ must vanish at any Brillouin zone edge. Note that Eqs. (13) and (14) jointly imply that the average velocity of all the electrons in any filled band must vanish exactly.

Near the bottom of any band with a minimum at $\underline{k} = 0$, such as the lowest band of Fig. 1(a), because of Eq. (14) a Taylor series around the minimum gives, to second order in $\underline{k}$,

$$E(\underline{k}) = E(0) + \sum_{i,j=1}^{3} A_{ij} k_i k_j \qquad (16)$$

where i and j represent the three possible spatial directions. It is always possible to choose a set of coordinate axes such that all $A_{ij} = 0$ for $i \neq j$. For such a set, we can define $m_i^* = \hbar^2/2A_{ii}$, and Eq. (16) becomes

$$E(\underline{k}) = E(0) + (\hbar^2/2) \left(\frac{k_x^2}{m_x^*} + \frac{k_y^2}{m_y^*} + \frac{k_z^2}{m_z^*} \right) \quad (17)$$

The quantities m_x^*, m_y^*, and m_z^* are known as the x, y, and z components of the effective mass. For cubic crystals, these three components must be equal for this type of a band, and we can write

$$E(k) = E_1 + \frac{\hbar^2 k^2}{2m} \quad (18)$$

where $m^* = m_x = m_y = m_z$ and $E_1 = E(0)$. Electrons in states in the band where Eq. (18) is a good approximation act just like free electrons, except for an altered mass m^*. The important point here is that these are the states available to electrons in the crystal, with the effects of the ion cores already taken into account.

Equations (16) through (18) hold only near the bottom of the band, and they cannot be expected to remain valid as the energy increases. We can always define an effective mass from the expression

$$m_{ij}^*(\underline{k}) = \hbar^2 \left[\frac{\partial^2 E(\underline{k})}{\partial k_i \partial k_j} \right]^{-1} \quad (19)$$

although when m^* is highly anisotropic and depends strongly on $\underline{k}$, it is not of much practical use. But no matter how complicated the band structure in a crystal is, we can always make use of the density-of-states function $g(E)$ defined as

$$g(E) = \frac{d\eta(E)}{dE} \quad (20)$$

where $\eta(E)$ is the number of states in the band per unit volume with energy less than E. Since the total number of states in any band per unit volume is 2N,

$$\int_{E_1}^{E_2} dE\, g(E) = 2N \tag{21}$$

where E_1 is the energy of the bottom of the band and E_2 is the energy of the top of the band. A typical density-of-states curve for a crystal is shown in Fig. 2. For the band of Eq. (18),

$$g(E) = \frac{1}{2\pi^2}\left(\frac{2m^*}{\hbar^2}\right)^{3/2}(E - E_1)^{1/2} \tag{22}$$

Such a relation is always valid near the band extremities. Since Eq. (22) represents a parabola, the range in which it remains appropriate is called the parabolic region of the band.

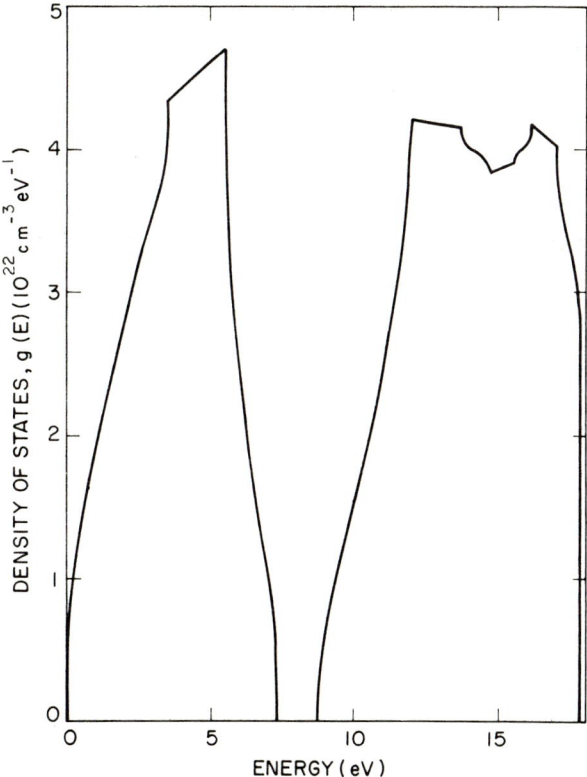

FIG. 2. Typical density of states as a function of energy for a periodic crystal. The sharp structure is due to three-dimensional periodicity.

Another useful parameter is the bandwidth Δ, defined as

$$\Delta \equiv E_2 - E_1 \qquad (23)$$

The bandwidth, effective mass, and density of states are closely related. If Δ is small, $g(E)$ must be large, as is seen from Eq. (21). Equation (22) then shows that m* is large. It is clear from Eq. (18) that $\nabla_{\underline{k}} E(\underline{k})$ must thus be small, and Eq. (13) indicates that the velocity of an electron in such a band is also small. Finally, Eq. (4) shows that such electrons have a low mobility.

D. ELECTRON STATISTICS

In Section II,C we described the states available to electrons as they move through a crystal. The ground state is the one in which all the electrons are in the lowest possible states. This becomes nontrivial only because of the Pauli exclusion principle, which prevents two electrons of the same spin from simultaneously occupying the same state. This principle is a general quantum mechanical result for any particle with half-integral spin. Thus the ground state is obtained by placing all the electrons in the crystal into the lowest states, one of each spin in each state, until we run out of electrons. This is the actual electronic state, however, only at T = 0. At finite temperatures some electrons will be excited into higher states. Thermodynamic equilibrium occurs when the electrons are distributed so that the probability that a state of energy E is occupied is

$$f_0(E) = \left(1 + \exp\frac{E - E_F}{kT}\right)^{-1} \qquad (24)$$

where E_F is a normalization parameter known as the Fermi energy. E_F is chosen so that n, the total density of electrons in the crystal, remains constant. Clearly we must have

$$n = \int_{-\infty}^{\infty} dE\, g(E) f_0(E) \qquad (25)$$

an expression that can be solved for E_F at any temperature. The function $f_0(E)$ in Eq. (24) is called the Fermi-Dirac distribution function [8]. It is essentially unity for $E \ll E_F$, and smoothly decreases from 1 to 0 in a range approximately $2kT$ on either side of E_F.

E. PHONONS AND LARGE POLARONS

The adiabatic approximation, discussed in Section II,B, enabled us to separate the problems of ionic motion and electronic motion, and also to solve the electronic problem for the single case of the ion cores in their (periodic) equilibrium positions. Actually, the ion cores vibrate around their equilibrium positions, and such lattice excitations increase the energy of the crystal. In the harmonic approximation, in which the restoring force on any ion away from its equilibrium position is treated only to first order, the lattice-vibrational spectrum is that of N coupled harmonic oscillators. The solutions to such a problem can be separated in terms of the (harmonic) normal modes of vibration, and any mode has the well known quantum mechanical spectrum given by

$$\epsilon_p = (p + \frac{1}{2}) \hbar\omega \qquad (26)$$

where p is any nonnegative integer and ω is the oscillator frequency. Since energy differences are integral multiples of $\hbar\omega$, we can think of the vibrational spectrum in terms of excitation of imaginary particles, called phonons. When p phonons in a given mode are excited, the additional energy of the crystal is $p\hbar\omega$. Note that there is always a minimum vibrational energy of $\hbar\omega/2$ for each mode, the zero-point energy, which is always required in quantum mechanical systems, since we cannot completely prevent an oscillator from vibrating via the uncertainty principle.

If the adiabatic approximation were exact, then phonons would not be of interest in relation to the electrical conductivity, since they would be completely decoupled from electrons. But actually electron-phonon interactions exist, and this means that electrons in

Bloch states can exchange energy and momentum with the lattice. We say that electrons can be scattered by phonons. Two types of processes are possible: (1) An electron can lose energy $\hbar\omega$ to the lattice by creating a phonon in the appropriate mode, provided an electronic state of energy $\hbar\omega$ below the original energy level is unoccupied (to avoid violation of the Pauli Exclusion Principle); and (2) an electron can gain energy $\hbar\omega$ from the lattice by absorbing the appropriate phonon, provided the higher electronic energy state is empty and at least one phonon in the correct mode is available. Phonons are particles that obey Bose-Einstein, rather than Fermi-Dirac, statistics, the distribution function taking the form

$$f(E) = (\exp \frac{\hbar\omega}{kT} - 1)^{-1} \tag{27}$$

Thus, only at high temperature will there be a sufficiently high probability of the higher-energy phonon being available to be absorbed by an electron. However, phonons can always be emitted if lower electronic states are unoccupied, since no Exclusion Principle exists for Bose-Einstein particles.

In ionic crystals, it is easy to see that the electron-phonon interaction is particularly strong. Consider an electron propagating through a crystal in a Bloch state. It always carries a negative charge, and so tends to attract positive ions toward it and to repel negative ions away from it. Thus it induces lattice vibrations as it moves, in particular the vibrations in which positive and negative ions move in opposite directions (the so-called optical modes). This interaction, arising from the strong Coulomb potential, is so strong that the adiabatic approximation breaks down. Rather than solve the pure electronic problem, we must, from the very beginning, solve for the states available to an electron together with its associated lattice vibrations. These states are then no longer simple electron states, and are called polaron states, the electron together with its associated phonons being thought of as a pseudoparticle, the polaron. In ordinary circumstances this pseudoparticle is a large polaron, the associated

ELECTRICAL TRANSPORT: GENERAL CONCEPTS 17

lattice deformation extending at least several interionic distances from the original electron. The only two modifications of the band theory discussed in Section II,C are (1) that the electronic effective mass is increased (due to the necessity of the electron always carrying its associated phonons with it as it moves), and (2) that the electronic energy is lowered (since the associated phonons induce a polarization of the positive and negative ions that attracts the negatively charged electron). Instead of electronic bands, we now have large-polaron bands, but with similar transport properties.

F. CORRELATIONS, MOTT TRANSITIONS, AND SMALL POLARONS

In Section II,E, we considered the situation when the adiabatic approximation breaks down. In the present section, we discuss the more serious consequences of the breakdown of the one-electron approximation. The one-electron approximation neglects electronic correlations, which represent the ability of two propagating electrons to move in such a way that they avoid approaching and thus mutually repelling each other. The Pauli exclusion principle represents strong correlations between electrons of parallel spin. By absolutely forbidding two such electrons to simultaneously occupy the same state, we force them to avoid each other. But in the one-electron approximation, nothing prevents two electrons of opposite spin from moving in the same region simultaneously. This comes about because the approximation requires that any electron feels only the average Coulomb repulsion from all the other electrons, and thus the other electrons cannot correlate their motion to avoid the electron under consideration.

There are examples in which it is obvious that neglect of correlations is a serious error. Consider a lattice of N widely spaced hydrogen atoms. Each hydrogen atom has three possible electronic states. A total of either 0, 1, or 2 electrons can be present around the hydrogen nucleus (proton), corresponding to H^+, H, or H^-, respectively. The respective ground-state energies are 0, -13.6 eV, and -14.4 eV, well known from atomic physics. But there must be exactly N total electrons on all the protons, to maintain charge neutrality. The ground state is clearly the one in which exactly one electron is on each proton; this has an energy of $E_0 = -(13.6 \text{ eV})N$. If we remove one electron from the vicinity of one proton and place it on another, the energy increases by 12.8 eV. If we were to solve this problem by means of the one-electron approximation, we would calculate a total energy of $E_{oE} = -(10.4 \text{ eV})N$. The neglect of correlations implies that each electron must be placed in one of the two possible states on each proton <u>at random</u>. Thus one-quarter of the protons will be unoccupied, one-half singly occupied, and one-quarter doubly occupied, producing the large energy quoted above. The one-electron energy is too high by the enormous value of 3.2 eV per atom, solely because of the neglect of correlations.

As we bring the isolated atoms together, the Coulomb repulsion between two electrons in the same vicinity is reduced by screening effects, due to polarization of all the other electrons. But bringing the atoms closer together results in an increasing bandwidth Δ. Thus a measure of the validity of the one-electron approximation is the bandwidth (or, alternatively, the interatomic spacing). Mott [9] presented physical arguments which showed that, for every material, a critical bandwidth exists below which the one-electron approximation sharply breaks down. In such a crystal, Bloch states are not the best electronic states, but a better approximation begins with states in which each electron is localized around a particular atom or ion. The transition from localized to Bloch states is called a <u>Mott transition</u>.

When an extra electron is localized in the vicinity of a particular ion in an ionic crystal, a local lattice deformation is induced. The electron attracts the positive ions in its vicinity toward its position, and repels the nearby negative ions. This induced deformation acts like an attractive potential, which can bind the electron so that it no longer can move freely through the lattice. As in Section II,E it is more accurate to consider states of the electron together with its associated local lattice distortion, a pseudoparticle which, in this case, is called a small polaron. As we shall see, the transport properties of small and large polarons are very different.

III. BOLTZMANN TRANSPORT THEORY

A. ELECTRONIC STATES IN A CRYSTAL IN THE PRESENCE OF AN APPLIED ELECTRIC FIELD

Given the validity of the adiabatic and one-electron approximations, we have found the electronic states in a crystal in the absence of an applied field. In order to discuss electrical conductivity, however, we must turn on a field and measure a current. Thus, we must also determine the dynamics of electrons in Bloch states in the presence of an applied electric field.

The real problem is extremely difficult for two reasons. First, the field destroys the equilibrium of the system, and energy is no longer conserved. Electrons can gain energy from the field, and we can no longer determine the average occupancy of a state by using the statistical distribution function given in Eq. (24). But more important, the Bloch states are no longer the actual electronic states of the problem. In the absence of a field, we have seen that the electronic states can be labeled by a wave vector $\underline{k}$, but that any electron is equally likely to be in the vicinity of any equivalent ion in the crystal. However, once a field is applied (negatively charged) electrons are accelerated opposite to the direction of the field, and clearly the proper states must have a strong dependence on position. Of course, since the external field does not have the periodicity of the crystal, the proof that electronic states are delocalized is no longer valid.

In general, we can hope to solve neither for the actual electronic states nor the real distribution function. However, one simplification is provided by the fact that the electrical conductivity is defined, as in Eq. (7), in the limit of small external fields, and in this limit perturbation theory is applicable. Thus, to a first approximation, electrons can be considered to move in Bloch states in which each electron acts like a free particle with an effective mass m^*. In this limit, it can also be shown [10] that we can usually treat the electron as a <u>classical</u> particle, specifying <u>both</u> its average position r and its k value. With these simplifications the transport problem can be handled.

We begin with an electron in the Bloch state $\underline{k}$. In the presence of a field $\underline{\mathcal{E}}(\underline{r})$ the electron has a total energy of $E(\underline{k}) - e\varphi(\underline{r})$, where e is the magnitude of the electronic charge and $\varphi(\underline{r})$ is the electrostatic potential, defined such that $\underline{\mathcal{E}}(\underline{r}) = -\nabla\varphi(\underline{r})$. Thus, the electron continuously gains energy from the field. After a time dt the electron has increased its energy by an amount dE equal to the product of force and distance:

$$dE = -e\,\underline{\mathcal{E}}(\underline{r}) \cdot \underline{v}\,dt \qquad (28)$$

where $\underline{v}$ is the electronic velocity, given by Eq. (13). But we can look at this increase in energy as arising from a changing value of electronic wave vector $\underline{k}$, such transitions being induced by the breakdown of the periodicity of the crystal by the field. Since $dE = \nabla_k E(\underline{k}) \cdot d\underline{k}$, Eqs. (13) and (28) show that

$$\frac{d\underline{k}}{dt} = \frac{-e\underline{\mathcal{E}}}{\hbar} \qquad (29)$$

Thus, Bloch electrons change their wave vector in the presence of a field according to

$$\underline{k}(t) = \underline{k}(0) - \frac{e\underline{\mathcal{E}}t}{\hbar} \qquad (30)$$

For example, let us consider a band such as the lowest band of Fig. 1(a). The corresponding electronic velocities are given in Fig. 1(b). It is clear from Eq. (30) that any (negatively charged) electron in a state near the bottom of the band is accelerated opposite to the direction of the applied field, thus giving a positive contribution to the conductivity. On the other hand, an electron in a state near the top of the band is accelerated parallel to the field, resulting in a negative contribution to the conductivity.*

One important result is clear already. If all the states in a given band are completely filled, the positive and negative contributions exactly cancel, and the resulting conductivity always vanishes. Thus a filled band never contributes to transport. Furthermore, it appears that if we wait for a sufficiently long time the average conductivity of an electron initially in any state k will be zero, since Eq. (30) implies that the electron will momentarily be in every state in the band for the same amount of time. Thus, in the one-dimensional example, after any multiple of the time $2\pi/e\mathcal{E}a$, the total contribution to the conductivity must vanish. This result, however, has no practical significance, for the simple reason that it is strictly true only for perfectly periodic crystals. But impurities, defects and lattice vibrations destroy perfect periodicity, and thus electrons do not remain in a given $\underline{k}$ state for a very long time, even in the absence of an external field. We can treat the effects of imperfections in terms of the scattering of electrons from one $\underline{k}$ state to another. The exact form of the scattering is largely irrelevant; what is important is that the scattering strongly tends to reduce the energy of the electron whenever lower-energy states are unoccupied. The reason for this is that the field has produced a nonequilibrium distribution of electrons, and scattering must always be such as to tend to restore equilibrium. Since the electrons have the excess energy that they gained from the field, they must lose this energy to the lattice via scattering processes. This is the mechanism for the I^2R Joule heating that always occurs when a current flows. But, more important, as long as the average time between scattering events is small

---

*Crystal periodicity implies that two points on opposite Brillouin zone edges, such as $k = -\pi/a$ and $k = \pi/a$ in the one-dimensional example of Fig. 1, must be taken to be identical. This leads to the phenomenon of Bragg reflection, the existence of which is intimately connected with the appearance of energy gaps [11].

compared to $2\pi/e\mathcal{E}a$, the contribution to conductivity of an electron initially anywhere in the band except very near the top will be positive. For small fields, $2\pi/e\mathcal{E}a$ is always sufficiently large for this to be the case. Since the electrons in a partially filled band are not initially near the top of the band, they all make a positive contribution to conductivity. Thus, any material containing a partially filled electronic band must be highly conducting.

Before the development of the band theory of solids, the "mystery of the unscattered electrons" was a major problem. It had been expected that electrons would be significantly scattered by all the ion cores, and thus that the average distance between scattering events (mean free path) should always be of the order of the interionic separation. Actually, values of the mean free path in excess of hundreds of interionic spacings were commonly observed. As we have seen, the band theory of solids shows that electrons in a periodic crystal, after the interactions with the ion cores are explicitly taken into account, still act just like free electrons, except for a modified mass. Thus only deviations from periodicity give rise to scattering, and very large mean free paths can be expected in near-perfect crystals at low temperatures.

B. BOLTZMANN EQUATION

In Section III,A we discussed the motion of a single electron in a crystal in the presence of a small applied electric field. All that remains is to take account of the large number of electrons present in a real solid. We have seen that in equilibrium electrons obey Fermi-Dirac statistics, characterized by the distribution function of Eq. (24). However, when an electric field is applied, equilibrium is destroyed, and f_0 no longer gives the average occupancy of a state, even if we use the correct energy $E = E(\underline{k}) - e\varphi(\underline{r})$.

In general, we can always define an actual distribution function $f(\underline{k},\underline{r},t)$ which gives the average occupancy at time t of the states where electrons are centered with positions between $\underline{r}$ and $\underline{r} + d\underline{r}$ and

ELECTRICAL TRANSPORT: GENERAL CONCEPTS

have wave vectors between $\underline{k}$ and $\underline{k} + d\underline{k}$. It is clear that we are relying on the classical description of an electron in order to define $f(\underline{k},\underline{r},t)$. This distribution function is the nonequilibrium analog of the Fermi-Dirac function $f_0(\underline{k},\underline{r},t)$ and, once it is known, it can be used to discuss all the transport properties of a crystal. Thus, transport theory is reduced to finding and solving an equation for f.

For a perfectly periodic system, no scattering occurs, and we can assume that the large number of electrons in the solid move through real space and wave-vector space in a continuous manner. An electron in a small volume element around the point $\underline{k}$, $\underline{r}$ at time t can leave it for one of two reasons: either because its velocity carries it away from $\underline{r}$, or because its rate of change of wave vector, given by Eq. (29), carries it away from $\underline{k}$. In a real system it can also be scattered away from $\underline{k}$ due to crystal imperfections. Assuming that the steady state has been reached, we can write down a continuity equation:

$$\nabla_{\underline{k}} f \cdot \frac{d\underline{k}}{dt} + \nabla_{\underline{r}} f \cdot \frac{d\underline{r}}{dt} = \left(\frac{\partial f}{\partial t}\right)_s \qquad (31)$$

where $(\partial f/\partial t)_s$ is the net rate of increase of f due to scattering processes. Equation (31) is called the Boltzmann equation. Its solutions can be used to analyze the transport properties of any material for which the assumptions underlying its derivation are applicable.

It is clear that the scattering term in the Boltzmann equation critically depends on f itself; in fact, it can be expressed as an integral over f. Thus Eq. (31) is an integrodifferential equation, which can be solved only by approximation techniques. The method most commonly used is the relaxation-time approximation. We assume that the effect of scattering processes is always to restore a local equilibrium situation described by the Fermi-Dirac distribution function $f_0(\underline{k},\underline{r},t)$ given by Eq. (24). We further assume that, if the electronic distribution is disturbed from local equilibrium, so that $f(\underline{k},\underline{r},t)$ differs from $f_0(\underline{k},\underline{r},t)$, then the effect of scattering is to restore f to f_0 such that the rate $(\partial f/\partial t)_s$ is proportional to the deviation from equilibrium:

$$\left(\frac{\partial f}{\partial t}\right)_s = -\frac{1}{\tau(\underline{k})}(f - f_0) \tag{32}$$

where $\tau(\underline{k})$ is a constant with the dimensions of time. Clearly τ must be of the order of magnitude of the time between scattering events; it is called the <u>relaxation time</u>. We can immediately integrate Eq. (32) to obtain

$$f(t) = f_0 + [f(0) - f_0] e^{-t/\tau} \tag{33}$$

Thus, deviations from equilibrium decay away exponentially in the relaxation-time approximation. Substituting Eq. (32) into (31), the Boltzmann equation, we find

$$\underline{v} \cdot \nabla f + \frac{d\underline{k}}{dt} \cdot \nabla_{\underline{k}} f = \frac{f - f_0}{\tau} \tag{34}$$

This is just a linear partial differential equation, which can be solved in many cases.

If the applied field is small, as we have assumed, we might expect only small deviations of f from f_0. Since Eq. (24) shows that f_0 depends only on the energy

$$E = E(\underline{k}) - e\varphi \tag{35}$$

then for local equilibrium the first term of Eq. (34) becomes

$$\underline{v} \cdot \nabla f_0 = \underline{v} \cdot \left(\frac{df_0}{dE}\right)(\nabla E)$$

$$= -e\left(\frac{df_0}{dE}\right)\underline{v} \cdot (\nabla \varphi)$$

$$= e\left(\frac{df_0}{dE}\right)\underline{v} \cdot \underline{\mathcal{E}} \tag{36}$$

Using Eq. (29) we obtain for the second term in Eq. (34)

ELECTRICAL TRANSPORT: GENERAL CONCEPTS

$$\frac{d\underline{k}}{dt} \cdot \nabla_k f_0 = -\frac{e}{\hbar} \cdot \left(\frac{df_0}{dE}\right) [\nabla_{\underline{k}} E(\underline{k})]$$

$$= -e \left(\frac{df_0}{dE}\right) \underline{v} \cdot \mathcal{E} \qquad (37)$$

where we have used Eq. (13). Thus Eq. (34) is identically satisfied at equilibrium.

In order to investigate the electrical conductivity, we apply a small, uniform electric field $\mathcal{E}$ in the z direction. Since the field is uniform, $\nabla f = 0$, and Eq. (34) is simply

$$-\frac{e}{\hbar} \left(\frac{\partial f}{\partial k_z}\right) = -\frac{f - f_0}{\tau(\underline{k})} \qquad (38)$$

If $\mathcal{E}$ is small, f and f_0 differ only slightly and we can write

$$f = f_0 + f_1 \qquad (39)$$

where $f_1 \ll f_0$. Then Eq. (38) becomes

$$\frac{e}{\hbar} \left(\frac{\partial f_0}{\partial k_z}\right) = \frac{f_1}{\tau} \qquad (40)$$

where we have neglected the second-order term $\partial f_1 / \partial k_z$. Since

$$\frac{\partial f_0}{\partial k_z} = \left(\frac{df_0}{dE}\right) \left(\frac{\partial E}{\partial k_z}\right)$$

$$= \hbar v_z \left(\frac{\partial f_0}{\partial E}\right) \qquad (41)$$

by Eq. (13), then Eq. (40) is trivially solved for f_1:

$$f_1 = e\mathcal{E}\tau v_z \left(\frac{df_0}{dE}\right) \qquad (42)$$

Equation (42) is the solution of the Boltzmann equation in the relaxation-time approximation.

From Eqs. (4) and (6) the conductivity can be written as

$$\sigma = -\frac{n\,e<v_z>}{\mathcal{E}}$$

$$= -\frac{ne}{\mathcal{E}} \frac{\int d\underline{k}\ v_z\ f(\underline{k},\underline{r},t)}{\int d\underline{k}\ f(\underline{k},\underline{r},t)} \qquad (43)$$

Substituting Eqs. (39) and (42) into (43), we find, to first order,

$$\sigma = -n\,e^2 \frac{\int d\underline{k}\ v_z^2 (df_0/dE)\ \tau}{\int d\underline{k}\ f_0} \qquad (44)$$

since

$$\int d\underline{k}\ v_z\ f_0 = \int d\underline{k}\ (\nabla_{\underline{k}} E)\ \frac{f_0}{\hbar} = 0$$

by virtue of f_0 being an even function of k [see Eqs. (14) and (24)] and $\nabla_{\underline{k}} E$ being an odd function of $\underline{k}$ [from Eq. (14)]. Physically this term clearly must vanish, since no current can flow at equilibrium.

Equation (44) can be evaluated immediately for the case of a metal, with a partially filled band of the form

$$E(\underline{k}) = E_0 + \frac{\hbar^2 k^2}{2m^*} \qquad (45)$$

From Eq. (24) we can see that (df_0/dE) is negative and significantly differs from zero only for $E \simeq E_F$. This means that provided τ does not depend on the direction of $\underline{k}$ it can be pulled out of the integral of Eq. (44) as τ_F, the value of the relaxation time at the Fermi surface. The numerator of Eq. (44) can be integrated by parts to obtain

$$\int d\underline{k}\ v_z^2 \left(\frac{df_0}{dE}\right) = \hbar^{\frac{1}{2}} \int d\underline{k}\ \left(\frac{\partial E}{\partial k_z}\right)\left(\frac{\partial E}{\partial k_z}\right)\left(\frac{df_0}{dE}\right)$$

$$= \hbar^{\frac{1}{2}} \int d\underline{k}\ \left(\frac{\partial E}{\partial k_z}\right)\left(\frac{\partial f_0}{\partial k_z}\right)$$

$$= -\hbar^{\frac{1}{2}} \int d\underline{k}\ f_0 \left(\frac{\partial^2 E}{\partial k_z^2}\right) \qquad (46)$$

ELECTRICAL TRANSPORT: GENERAL CONCEPTS

But from Eq. (45)

$$\frac{\partial^2 E}{\partial k_z^2} = \frac{\hbar^2}{m^*} \qquad (47)$$

Substituting Eqs. (46) and (47) into (44), we finally obtain

$$\sigma = \frac{ne^2 \tau_F}{m^*} \qquad (48)$$

This important result could have been obtained very simply, essentially by dimensional analysis, but its derivation by means of the Boltzmann equation shows its general applicability. Note that all electrons in a partially filled band contribute to electrical conductivity, not only the electrons near the Fermi energy.

The Boltzmann equation can be used to evaluate the conductivity of a semiconductor as well, although a different approximation must be used for integrating over $\underline{k}$ [10, pp. 180-189].

The theoretical expression for conductivity, for either metals or semiconductors, is incomplete unless the relaxation time can be calculated. Several scattering mechanisms have been analyzed [12], including scattering by grain boundaries, stacking faults, dislocations, vacancies, interstitials, compositional disorder, spin disorder, lattice vibrations, ionized impurities, neutral impurities, and electron-electron interactions. In most of these cases, expressions for τ have been derived and compared with experiment. When simultaneous scattering mechanisms are important, the problem is much like that of a series circuit. The expressions for the reciprocals of the relaxation times are added to give a total reciprocal τ:

$$\frac{1}{\tau} = \sum_i \frac{1}{\tau_i} \qquad (49)$$

where τ_i is the relaxation time for scattering mechanism i. This expression just represents the fact that any one of several types of interaction can scatter an electron as it moves through the crystal. It is interesting to note that even in amorphous materials the effects of long-range positional disorder can be handled in terms of a scattering approach based on the relaxation-time approximation.

In ionic crystals for which large polaron theory is applicable, Boltzmann transport theory usually remains valid provided the polaron effective mass is used for m^*. But we must bear in mind that large polarons are pseudoparticles which must have a finite lifetime, and the Boltzmann approach cannot be expected to apply if the lifetime is shorter than the relaxation time. However, this is not generally a problem if polaron theory applies.

C. HOPPING CONDUCTION

As we discussed in Section II,F, localized electrons in ionic Mott insulators induce lattice deformations, which trap the would-be carriers at the position of particular ion cores. Once trapped, such an electron will no longer participate in electrical conduction. However, in the course of a random lattice vibration it is possible that the environment of the ion core at which the electron is trapped momentarily becomes identical with that of a nearest-neighbor ion core. At that time the electron can, without further expenditure of energy, hop from the former site to the latter. This can be considered a diffusion of the trapped electron through the crystal. In the presence of a small applied field, preferential diffusion of the negatively charged particles opposite to the direction of the field produces a positive contribution to conductivity. This mechanism is known as <u>thermally</u> <u>activated</u> <u>hopping</u> <u>conduction</u>.

The major characteristic of hopping conduction is the strong temperature dependence of the associated mobility. This arises because at low temperatures, the lattice vibrates sluggishly, and the chances of obtaining two nearest-neighbor equivalent sites are very small. However, at high temperatures the lattice vibrations are vigorous and the hopping probability is exponentially larger. Thus hopping conduction is in principle distinguishable from band conduction by means of mobility measurements.

As previously noted, conduction in ionic Mott insulators is more accurately discussed in pseudoparticle language, in which small-polaron states, rather than electronic states, are fundamental. It can be shown [13] that at low temperatures the periodic zero-point lattice vibrations are sufficient to allow a bandlike mechanism of small polaron conductivity, but above approximately half the Debye temperature small polarons conduct only by hopping--the electron together with its associated lattice distortion hops through the crystal.

Small-polaron mobility can be analyzed by means of the Boltzmann equation. It is clear that self trapping cannot occur unless the hopping time τ is large compared to the lattice vibration period τ_0. This relation gives an upper limit for the hopping mobility of 0.1 cm^2/V-sec. A lower limit for bandlike mobility can be derived from the condition that the mean free path, which is of the order of the product of τ and the thermal velocity, be greater than the interatomic separation a. This condition leads to a lower limit of $5(m/m^*)^{1/2}$ cm^2/V-sec for mobility in a Bloch band. For a large effective mass, say $m^*/m \simeq 25$, this lower limit is 1 cm^2/V-sec. Thus direct mobility measurements at one temperature can distinguish band from hopping conduction. However, a direct measurement of mobility is often more difficult than obtaining the variation of mobility with temperature.

D. IONIC CONDUCTION

Any charged particle that can be accelerated by an electric field can participate in conduction. Thus we might expect, for example, that the positive ion cores in a metal or covalent semiconductor contribute to the observed conductivity. This contribution is completely negligible, however, for two reasons. First, it is evident from a comparison of Eqs. (11) and (12) that the mobilities

of the ion cores are at least a factor of 10^3 lower than those of
the electrons, even when the electrons are localized with a mean
free path of the order of a lattice parameter. Ionic mobilities
rarely exceed the order of 10^{-3} cm^2/V-sec, extremely small compared
with electronic mobilities, which are of the order of 10-10^6 cm^2/V-
sec. Second, ions can conduct only by means of diffusion through the
lattice, and such diffusion requires the existence of either vacancies or interstitials. It takes an activation energy of approximately 1-5 eV to create an interstitial-vacancy pair and an energy of
equivalent magnitude to move the vacancy or interstitial ion through
the material. The total activation energy for intrinsic ionic conduction in this case is half the energy of creation of a defect plus
the energy of motion, i.e., of the order of 5 eV. Thus, the contribution to ionic conductivity at room temperature (kT $\simeq$ 0.025 eV) may
be as small as

$$\sigma_I \simeq N \exp\left(\frac{-5 \text{ eV}}{kT}\right) e\mu_I$$

$$\simeq 10^{-86} \text{ ohm}^{-1} \text{ cm}^{-1} \tag{50}$$

Even at temperatures of 2000°K, $\sigma_I \simeq 10^{-12}$ ohm^{-1} cm^{-1}. Clearly, if
any electronic conduction is possible, the contribution of the ion
cores to conductivity is negligible.

On the other hand, for ionic crystals such as NaCl, the energy
necessary to free an electron from one of the closed-shell configurations and place it on a distant ion could well exceed the total
activation energy necessary to create and move a defect. Then, at
least at low temperatures, ionic conduction will dominate.

Whenever parallel mechanisms of conduction exist, the total
conductivity is the sum of the individual contributions. For semiconductors and insulators, we can express the total conductivity as

$$\sigma = \sum_i \sigma_i \exp\left(\frac{-E_i}{kT}\right) \tag{51}$$

Clearly the process with the highest σ_i always predominates at sufficiently high temperatures, while the process with the lowest E_i must predominate at sufficiently low temperatures. Other processes either are observable in an intermediate temperature range or are completely unobservable, depending on the values of their pre-exponentials σ_i and their activation energies E_i. The situation is as shown in Fig. 3. Note that simultaneous conduction mechanisms act as a parallel circuit, although simultaneous scattering mechanisms act as a series circuit. This contrast is evident in Eqs. (49) and (51).

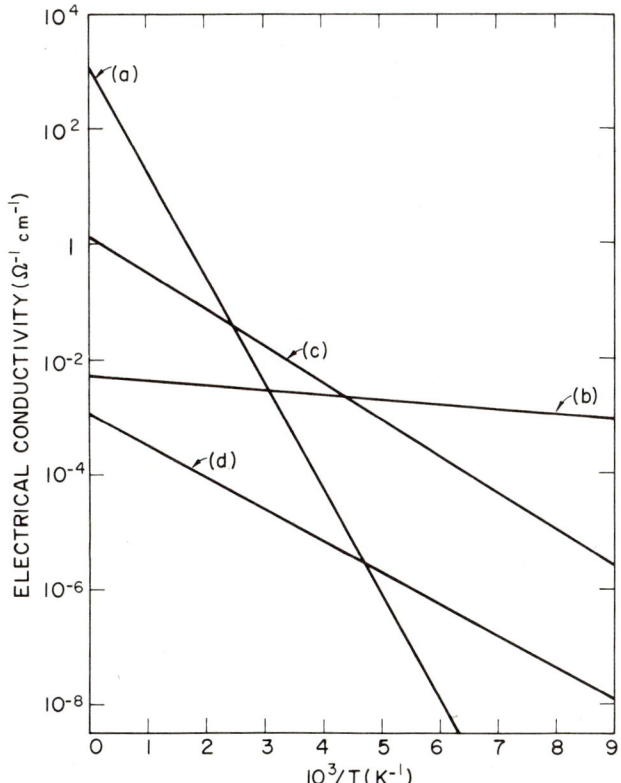

Fig. 3. Plot of logarithm of the conductivity as a function of reciprocal temperature for each of several parallel conduction mechanisms. Process (a) predominates at high temperatures; process (b) predominates at low temperatures; process (c) can be observed in a narrow temperature range; process (d) is completely unobservable.

IV. KUBO TRANSPORT THEORY

Boltzmann transport theory, discussed in Section III, depends on several important assumptions. First, the theory is semiclassical, in the sense that we specified both the position and wave vector of a given electron. Second, we assumed that the applied field is sufficiently small that perturbation theory, beginning with the equilibrium states, is applicable. Third, we tacitly assumed [14] that the deviations from crystalline periodicity are sufficiently small that the relaxation time, i.e., the time between electronic collisions, is long compared with the duration of a collision ($\sim \hbar/kT$). If this is not the case, successive scattering events cannot be independent, and the relaxation time approximation fails. It is this last assumption that leads to the downfall of Boltzmann theory, since it fails whenever strong scattering exists, such that $\tau < 10^{-14}$ sec at room temperature. In low-mobility materials, τ is always much smaller than this value, and the Boltzmann approach cannot be justified. However, even in metals the assumption $\tau \gg \hbar/kT$ fails at low temperatures, when strong impurity scattering remains, and at high temperatures, when strong scattering by lattice vibrations forces τ to be of the same magnitude as $\hbar/kT$.

To overcome this weakness in transport theory, Kubo [15] suggested an alternative approach based on the first-order response of the electrical current to a small applied field. A formal exact expression for the conductivity can be derived in terms of a complex correlation function:

$$\sigma_{ij} = \int_0^{1/kT} d\lambda \int_0^{\infty} dt < J_i(-i\hbar\lambda) \, J_j(t) > \tag{52}$$

where J_i is the current operator. Given the dynamics of any particular system, Eq. (52) provides a prescription for evaluation of the conductivity, even when the scattering is strong. Despite its formal nature, the physics behind the Kubo transport theory is straightforward. We simply look for the differential increment in electrical current resulting from the application of an infinitesimal pulsed electric field, and identify the conductivity via Eq. (2). Ultimately, just as indicated in the Introduction, electrical conductivity is the net transport of charge in the presence of an applied field, no matter how sophisticated the mathematical techniques used in its evaluation.

REFERENCES

1. See, e.g., W. Yourgrau, A. van der Merwe, and G. Raw, Treatise on Irreversible and Statistical Thermophysics, Macmillan, New York, 1966 for an excellent treatment.

2. See, e.g., W. Kauzmann, Quantum Chemistry, Academic Press, New York, 1957.

3. M. Born and J. R. Oppenheimer, Ann. Phys. (Leipzig), 84, 457 (1927).

4. J. R. Reitz, Solid State Phys., 1, 2 (1955).

5. These are called the Born-von Kármán boundary conditions. See M. Born and T. von Kármán, Phys. Z., 13, 297 (1912).

6. H. Jones, The Theory of Brillouin Zones and Electronic States in Crystals, North-Holland, Amsterdam, 1960.

7. See, e.g., G. Weinreich, Solids: Elementary Theory for Advanced Students, p. 111, Wiley, New York, 1965.

8. See E. A. Desloge, Statistical Physics, Holt, Rinehart and Winston, New York, 1966, for a derivation.

9. N. F. Mott, Phil. Mag., 6, 287 (1961).

10. A. C. Smith, J. F. Janak, and R. B. Adler, Electronic Conduction in Solids, pp. 121-125, McGraw-Hill, New York, 1967.

11. See, e.g., C. Kittel, Introduction to Solid State Physics, p. 256, Wiley, New York, 1966.

12. F. J. Blatt, Physics of Electronic Conduction in Solids, pp. 137-181, McGraw-Hill, New York, 1968.

13. T. Holstein, Ann. Phys. (New York), 8, 343 (1959).

14. R. E. Peierls, <u>Quantum Theory of Solids</u>, p. 124, Oxford Univ. Press, 1954.

15. R. Kubo, <u>J. Phys. Soc. Jap.</u>, <u>12</u>, 570 (1957).

Chapter 2

EXPERIMENTAL TECHNIQUES

Robert N. Blumenthal and Martin A. Seitz

Department of Mechanical Engineering
Marquette University
Milwaukee, Wisconsin

I.	INTRODUCTION	36
II.	GENERAL EXPERIMENTAL CONSIDERATIONS	38
	A. Sample Preparation	38
	B. Electroding Techniques	46
	C. Selection of Materials for Constructing Measuring Systems	57
III.	THERMODYNAMIC MEASUREMENTS	60
	A. Galvanic-Cell Measurements	61
	B. Gas-Equilibrium Measurements	76
IV.	ELECTRICAL CONDUCTIVITY	93
	A. Noise Considerations	93
	B. Experimental Methods	95
V.	TRANSFERENCE MEASUREMENTS OF MIXED CONDUCTORS	112
	A. Reversible Electrodes	113
	B. Polarization Measurements	131
VI.	THERMOELECTRIC MEASUREMENTS	143
VII.	MOBILITY	152
	REFERENCES	168

I. INTRODUCTION

The electrical conductivity of ceramic materials encompasses a wide range of values which characterize insulators, semiconductors, and metallic conductors. The electrical behavior is usually strongly dependent upon temperature and composition. For example, a small change in the composition may transform an insulating ceramic into a semiconductor. When the resistivity is high, conduction may occur by both electronic and ionic processes. The electrical conductivity of a homogeneous single-phase material is generally described by the following expression

$$\sigma = \sum_i n_i z_i \mu_i \qquad (1)$$

where n_i is the concentration of charge carriers with charge z_i and mobility μ_i. A complete characterization of the electrical conductivity requires that the charge, concentration and mobility of each charge carrier be determined as a function of the pertinent parameters (e.g., temperature, composition, atmosphere).

In general these quantities cannot be determined directly by experiment. Usually the results from several different experimental measurements must be used in combination with the theoretical interpretation in order to provide this information. These experimental methods usually include one or more of the following types of measurements: thermodynamic, electrical conductivity, transference number, thermoelectric, and mobility.

One of the most important factors that must be considered when making property measurements is the determination of the variables that must be specified in order to uniquely determine the state of the system (i.e., in order that the properties of the material have a uniquely defined value). In many applications the phase rule may be used to answer this question. The phase rule is given by the expression

EXPERIMENTAL TECHNIQUES

$$F = C - P + 2 \tag{2}$$

where F, the number of independent variables, is called the number of degrees of freedom, C is the number of components and P is the number of phases. For example, in a binary system such as TiO_{2-x} in equilibrium with its vapor, the number of degrees of freedom is 2. Thus, upon arbitrarily fixing both the temperature and the partial pressure of oxygen, the composition of TiO_{2-x} is uniquely determined, i.e., $x = f(T, P_{O_2})$. For a ternary system such as $BaTiO_{3-x}$, one additional degree of freedom must be controlled. This can be the cation ratio, Ba/Ti, or any other suitable compositional variable.

Since the electrical conductivity of a single-phase material depends upon the type, concentration, and mobility of the charge carriers, the electrical conductivity will be uniquely determined when the temperature and composition are fixed, provided the mobility is a function of temperature and composition only. This is generally true for single-crystal specimens if the effects of dislocations and impurities are negligible. The conductivity of polycrystalline ceramics, however, may differ from that of single crystals because of the influence of grain boundaries and pores. For a ceramic with a high electronic mobility this effect may arise from the additional scattering at the grain boundaries and pores. This effect may also occur if the grain boundaries possess a conductivity different from that of the grains or if significant conduction occurs via an electron gas in the pores. These problems are discussed in detail in Chapter 9 (Part B of this treatise).

The above considerations assume that the ceramic specimen is a single phase in thermodynamic equilibrium with its environment. At high temperatures, kinetic processes are sufficiently rapid that the specimen can equilibrate and the state of the system can be uniquely defined by controlling the temperature and the composition of the specimen. However, if measurements are made at low temperatures, where equilibrium with the environment cannot be achieved in a reasonable period of time, the concentration of defects characteristic of a high temperature may be frozen in and the past history of the specimen must be completely described.

Before proceeding to a detailed description of the aforementioned experimental methods a general discussion will be presented of the important factors and problems that the experimentalist should be cognizant of when making thermodynamic and electrical property measurements. These include the following considerations: preparation of the specimen, chemical and structural characterization of the specimen, techniques for electroding the specimen, and selection of materials for the construction of the measuring system.

II. GENERAL EXPERIMENTAL CONSIDERATIONS

A. SAMPLE PREPARATION

When attempting to characterize the properties of ceramic materials the nature of the starting materials should be the first subject of the investigator's attention. Thus information as to the purity, perfection, and availability of research materials is of prime importance. Cleland [1] has discussed the major sources of this type of information and these are presented in Table 1. Of the sources listed, the Research Materials Information Center seems to be the most comprehensive. A number of other specific references are also available in the literature [2-4]. For research purposes single crystals of high purity and perfection are often desirable. The method of crystal growth strongly influences the size and quality of single-crystal materials and should be considered when acquiring sample materials.

Screw dislocations are generally formed during the crystal growth process or are due to imperfections in the starting material; edge dislocations arise during cooling, if the crystals are grown in nonlinear or large thermal gradients [1-5]. Dislocations can also be introduced into single-crystal specimens during cutting, grinding,

TABLE 1

Sources of Information on Purity, Perfection,
and Availability of Research Materials

Source	Type of Information Available
Research Materials Information Center Bldg. 3001 Oak Ridge National Laboratory P.O. Box X Oak Ridge, Tennessee 37830	Information is available on the initial purification, crystal growth, final characterization and availability of a wide variety of materials of research quality.
Defense Ceramic Information Center Battelle Memorial Institute Columbus, Ohio	Information is available to those who qualify on materials composition, processing, fabrication, property and performance data, and fundamental aspects of processing and behavior.
Electronic Properties Information Center Air Force Systems Command Hughes Aircraft Company Culver City, California Defense Documentations Center Alexandria, Virginia	Electrical property data on Al_2O_3, BeO, MgO, CdO, PbO, and ZnO is available.
<u>Materials Handbook</u> by G. S. Brady McGraw-Hill Book Company	This source carries an up-to-date tabulation of a wide variety of materials. It is periodically updated and is currently in its 10th edition.
1. Materials Research Bulletin 2. Journal of Crystal Growth 3. Journal of Materials Science 4. Journal of the American Ceramic Society	These journals provide additional information on a periodic basis.

and lapping operations. These generally would be formed at the surface of the sample and could be removed using the proper etchant. In general, low-speed diamond saws and wire saws introduce less mechanical damage than the higher-speed cutting methods. Etch-pit and X-ray techniques have been discussed in the literature, if one is interested in characterizing the dislocation densities at the surface or within the bulk of a single-crystal ceramic specimen [9].

Deleterious chemical impurities can either be present in the starting materials or be introduced during the crystal growth process. The diversity of physical properties (e.g., melting point, boiling point, vapor pressure) between various ceramic materials dictates the use of a variety of crystal-growth techniques [5-9]. Crystal growth methods may be divided into two general categories, monocomponent and polycomponent. The term "monocomponent growth" implies that only the material composing the desired single crystal is involved in the growth reaction. Polycomponent growth techniques use foreign substances as carrier, reactant, or medium for growth. Alternatively, four different ways of growing single crystals, depending on the phase from which they grow, can be distinguished. These are growth from the vapor phase, from the liquid phase, from a supercritical or "fluid" phase, and from one or more solid phases. Factors that influence the method of crystal growth chosen for a particular ceramic material include size, chemical purity, and structural perfection of the desired crystal along with the ease of adding specific impurity dopants. A number of excellent reviews are available that discuss the attributes of the various methods of crystal growth [5-9].

The tensor properties of materials are generally of rank greater than zero, so the orientation of the major crystallographic axes within a ceramic sample and the point-group symmetry of the material should be of some concern to the investigator [10]. The Laue back-reflection method is the most often used technique to orient single-crystal specimens [11]. This method also aids in determining whether one has a single crystal in the first place.

EXPERIMENTAL TECHNIQUES

Polycrystalline samples may be used where the phenomena being studied are not strongly controlled by grain-boundary effects. Fabrication and doping techniques are normally easier when one is dealing with polycrystalline materials. The principal method of fabricating polycrystalline ceramic specimens is by sintering [5, 12,13]. Sintering is a process in which an agglomerate of fine particles, when baked at an elevated temperature, forms a strong, dense body. Apparatus required to produce sintered samples is relatively simple and may be found in most laboratories.

Sintered specimens are normally formed to a specific geometry by pressing the powdered material in an appropriately designed die [12,14]. These dies can be constructed of hardened steel, chromium-plated steel, or other abrasive-resistant alloys. It is generally found that upon applying pressure to a sample in a die, there is a large initial increase in the bulk density. However, as the pressure is increased a point is reached where pressure has little additional influence. Therefore the pressure used should be sufficient to give the sample green strength or large enough to take advantage of the initial rate of increase in bulk density with pressure, but should not be excessive. If too high a pressure is used, flaws and cracks will result in the sample due to the nonuniform stresses that developed in the die. Typical pressures used range between 3000 and 10,000 psi. Organic binders (polyvinyl alcohol, dextrine, waxes, wax emulsives, acrylates, etc.), which can be driven off during the firing operation, are sometimes used when further green strength is desired.

Three stages occur during the sintering process [5,12,13]. In the first stage, the fine grains become smooth and begin to unite along their common boundaries to form a spongelike structure. In the second stage, the common surfaces between the grains grow, while the pores shrink, until a condition is reached where isolated pores exist at the intersections of various grains. In the final stage, the remaining pores are eliminated, while the average particle size increases, larger grains growing at the expense of smaller ones.

These three stages take place by viscous flow, evaporation-condensation, surface diffusion, or bulk diffusion.

The objective of sintering is to produce a specimen of maximum bulk density. Densities approaching 98% of theoretical density have been achieved by this method. The principal variables controlling the sintering process are initial particle size and size distribution, and temperature. Impurities and nonstoichiometric defects enhance the rate of diffusion and can also increase the sintering rate and final bulk density. It is found that the sintering rate steadily decreases with time, so that merely sintering for extensive periods of time to obtain improved properties becomes impractical. Particle size is important since the rate of material transferred via evaporation-condensation, and diffusion processes have an inverse-power-law dependence on the radius of the particles [5,12,13]. It is advantageous to have a range of particle size, since the smaller particles will tend to fill the interstice openings between the larger particles. Temperature, because of its influence on vapor pressure, diffusion, and plastic flow, becomes the prime variable when attempting to achieve high-density ceramics. Densification in sintering is sometimes limited because of extensive secondary grain growth. This produces many pores that become isolated from grain boundaries, which normally act as vacancy sinks. The diffusion distances between vacancies and grain boundaries become high and the rate of sintering is diminished.

In order to maintain sample purity it is advisable to surround the sample with powdered material of the same composition as the sample during the sintering process (see Fig. 1). This minimizes the introduction of impurities via diffusion from the boat and furnace chamber in which the sintering is taking place.

An alternative to sintering is the process of hot-pressing, which can lead to bulk densities approaching 100% of anticipated values [15]. Hot-pressing is similar to the sintering process in many respects, except that pressure is applied during the densification process. In this process powdered material is loaded into

EXPERIMENTAL TECHNIQUES   43

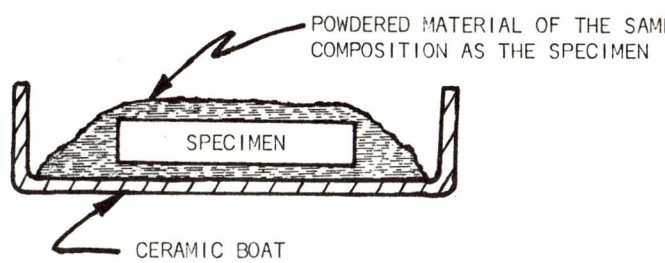

FIG. 1. A method of maintaining sample purity during sintering.

an appropriately designed die-and-plunger assembly, which is generally constructed of graphite. Pressures in the range of 1000-50,000 psi and temperatures as high as 2500°C have been used in hot-pressing processes. The application of pressure leads to particle rearrangement, enhanced diffusion, grain-boundary sliding, creep, and plastic flow. It is generally found that the application of pressure yields higher bulk densities at lower temperatures than pressureless sintering. Densification rate is found to increase with increasing temperature and pressure, and with decreasing particle size.

Property measurements sometimes require that ceramic materials be doped with a specific concentration of known impurities. In order to dope single crystals the desired impurity must be added either during the growth process or, for existing crystals, via diffusion from the external surfaces of the crystal. Although the first of these methods of doping is the most desirable, it requires the investigator to become involved in crystal growth if no readily available source of doped crystals exists. Diffusive doping can be used if the diffusion coefficient of the impurity dopant in the ceramic is large and the sample size is small. There is, however, the danger of a nonuniform impurity concentration within the final specimen.

Doped polycrystalline samples are easier to prepare than their single-crystal counterparts. Several methods are commonly used for this purpose [16]. The dopant can be dissolved in the melt of the host material if its melting point is not too high. Melting, however, may introduce errors due to vaporization losses or due to reactions with the crucible materials. Solid-state reaction can also be used to introduce an impurity dopant into a host material. Impurity cations are often added by mixing their carbonate with the powdered host material. Because of the difficulty encountered in achieving equilibrium, reactive and finely divided powder, which has been intimately mixed, must be homogenized using high sintering temperatures. A reactive and finely divided powder may sometimes be obtained by drying a nitrate mixture by evaporation, or coprecipitating the required metal cations as hydroxides or oxalates. If a suitable solvent for the host material and doping agent can be found, they might be dissolved and subsequently coprecipitated to yield a finely divided powder of the desired composition.

Foreign impurities can have a significant influence on the results of an investigation of ceramic materials. These impurities may be introduced intentionally, as discussed previously, or unintentionally during the fabrication and measurement processes. Surface contaminants, which can be introduced during handling, by contact with surrounding containers, or from dust and dirt in the work area, may diffuse into the interior of samples when subsequent high-temperature measurements are performed. It is, therefore, advisable to consider the various chemical analysis techniques that are available to the investigator to determine the concentration and distribution of foreign impurities. Excellent reviews of the various methods of chemical analysis are contained in the literature [5,9], a summary of which is given in Table 2. Because of the skill needed to obtain reliable chemical analyses, the experimenter must often seek out personnel and laboratories equipped for this aspect of his research.

Along with the above chemical characterization, X-ray powder pattern methods can be used to determine the predominant crystallographic phases or the phase distribution within a ceramic sample [11].

EXPERIMENTAL TECHNIQUES   45

TABLE 2

Methods of Chemical Analysis

Method	Remarks
Spectography	This method of analysis is often used for metals, but can be extended to ceramic materials. Without concentration and separation sensitivities of the order of 0.5-1 ppm have been reached, while with concentration and separation techniques, sensitivities approaching 0.01 ppm have been achieved. The method is moderately insensitive to the lighter elements such as O, N, H, P, S and Cl.
Mass spectroscopy	Mass spectroscopy can be used to detect essentially all elements. Typical sensitivities are in the order of 10^{-2}-1 ppm, although if stable isotopes are available the sensitivity can be increased to 10^{-4} ppm for the majority of the elements.
Tracer	If radioactive isotopes with a sufficiently long lifetime exist in the sample, or if a certain percentage of a particular element is known to exist as such an isotope, then sensitivities approaching 10^{-2}-10^{-7} ppm can be achieved by this method.
Neutron activation	This is a variation of the tracer technique. The sample is bombarded with neutrons which will transform many elements to radioactive isotopes. Upon counting, the energy of the pulses indicates the type of element and the intensity of its concentration. Sensitivities of 10^{-2}-1 ppm have been reached by this method.
Colorimetry and fluorometry	Using concentration and separation techniques, sensitivities of 10^{-1}-10^{1} ppm have been achieved. This method is useful for all impurities listed in the periodic table.
Polarography	By concentration many elements may be detected with a sensitivity approaching 10^{-1} ppm.

X-ray fluorescence	Upon electron bombardment or X-radiation, X-ray fluorescence is excited in the sample. Impurities can be identified via their characteristic fluorescent wavelength. Their concentration is proportional to the intensity of fluorescence. Typically, sensitivities are of the order of 10^2-10^3 ppm, but a sensitivity of 1-10 ppm is attainable using separation and concentration techniques. By the use of the electron-probe X-ray microanalyzer, concentration profiles and compositions of precipitates may be determined.
Other	Infrared absorptiometry, vacuum fusion, or methods based on some particular physical property (e.g., conductivity, photoconductivity, luminescence, parametric resonance) can also be used to determine varying concentrations of impurities within ceramic samples. Parametric resonance usually requires intimate knowledge of the material being investigated.

This is especially important when high doping levels are being used and there is a danger of precipitates forming within the sample.

B. ELECTRODING TECHNIQUES

In order to perform electrical measurements on materials, electrical contacts of high electrical, chemical, and mechanical quality under a wide range of temperatures and gas atmospheres must be applied to the specimens under investigation. In the discussion that follows an attempt will be made to develop a general understanding of the problems associated with producing high-quality electrical contacts and to review various methods available to accomplish this objective.

In measurements of the electrical conductivity of a material the electrical contact must not impede the flow of charge carriers. The flow of electronic carriers can be impeded by barrier layers and/or nonohmic contacts. If high-resistance surface layers existed between the electrode and the specimen, the applied electric field

EXPERIMENTAL TECHNIQUES   47

would concentrate at these layers. Therefore the applied field would not be representative of the field across the bulk specimen. These layers can be the result of air gaps [17] or a layer of foreign material between the electrode and the specimen, or of an inhomogeneous composition of the specimen in the vicinity of its surfaces [18,19]. (See Fig. 2) Nonohmic behavior of a contact is the result of a difference in work function between a metal contact (φ_m) and the ceramic material to which it is applied (φ_s). The thermodynamic requirement that Fermi energy of the specimen (E_{F_s}) and of its contact metal (E_{F_m}) must be of the same value after they unite can lead to the formation of an energy barrier. The magnitude of this energy barrier is equal to the difference in work function of the two materials. (See Fig. 3)

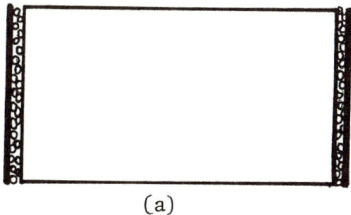

(a)

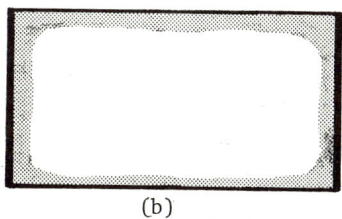

(b)

FIG. 2. (a) Specimen with a foreign substance between the electrode and sample. (b) Specimen with an inhomogeneous composition existing near its surfaces.

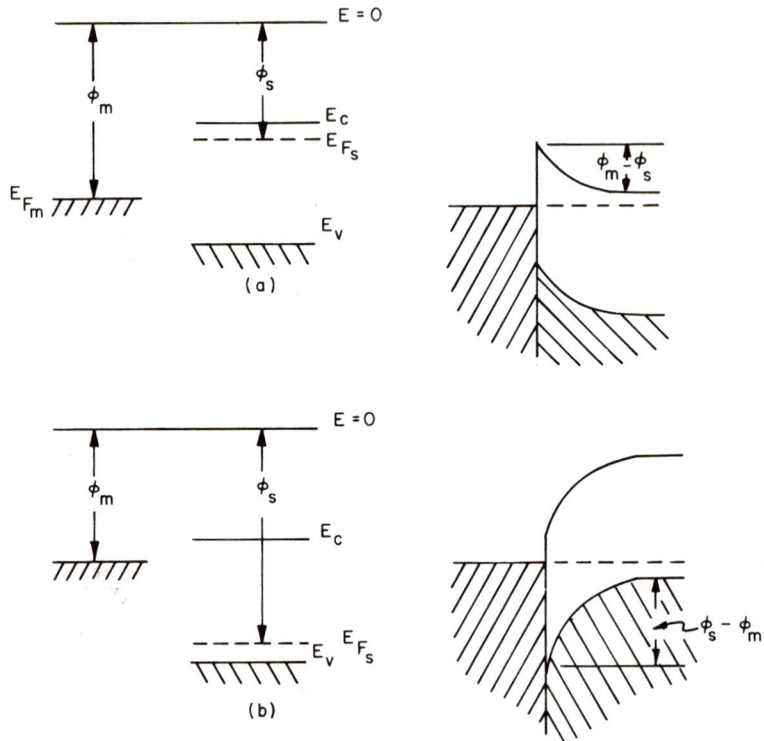

FIG. 3. (a) Band diagram of an n-type semiconductor and a metal, where $\varphi_m > \varphi_s$, before and after they are joined. (b) Band diagram of a p-type semiconductor and a metal, where $\varphi_m < \varphi_s$, before and after they are joined.

The problems associated with barrier layers can sometimes be eliminated by applying pressure to the contact area to minimize air gaps or by allowing the electrode material to diffuse a short distance into the interior of the specimen. Nonohmic behavior of a contact can often be avoided by choosing, as a contact material, a metal whose work function is smaller than that of an n-type specimen [20,21]. A second means of achieving ohmic behavior would be to choose electrode materials which would cause an n-type material to become more n-type, or a p-type material more p-type if they diffused into the specimen [20]. A high concentration of surface-energy states in the forbidden band gap of a material would act as an infinite source or sink to charge carriers [22]. Therefore, roughing the surface would aid in achieving an ohmic behavior of the contact.

The chemical aspects of the materials used for electrical contacts must also be considered when choosing an appropriate contact material. The electrode materials chosen must be chemically inert with respect to the specimen and must not appreciably diffuse into the specimen at the temperatures and ambient atmospheres used in the experiment. At low temperatures a wide variety of materials are available which fulfill these conditions, but as the temperature of measurement is increased these requirements become difficult to fulfill. Further difficulties may be encountered due to melting or vaporization of the electrode materials at elevated temperatures. A listing of commonly used electrode materials is given in Table 3.

If reliable electrical measurements are to be conducted, the contacts to the specimen must have sufficient mechanical strength to withstand a moderate amount of handling. In considering the mechanical strength of an electrical contact, two aspects must be explored. These are the bond between the material composing the electrode and the specimen; and the bond between the wires leading to the measuring instrumentations and the electrode material. Adhesive bonds between a metal and a ceramic material may be described either as mechanical or specific. Mechanical bonds may be formed by an interlocking or wedging of the metal electrode material and the ceramic specimen. The rough surface of a polycrystalline specimen would provide crevices by which this interlocking could take place. Surface roughness and crevices would have to be created on smooth surfaces by grinding or sanding. Specific adhesion depends on chemical-physical bonds, i.e., polar, covalent, and van der Waals. Bonds of these types exist when an actual wetting of the ceramic surface occurs. Other factors that influence the mechanical-electrical behavior of an electrical contact would be the nature of

TABLE 3

Commonly Used Electrode Materials

Element	m.p. (°C)	Remarks
Hg	-38.87	Mercury can be used as a liquid metal contact at low temperatures.
Ga	+29.78	Gallium is used as an alloying agent to form low-temperature contacts.
In	156.4	These materials can be used as low-temperature electrodes either by themselves or upon alloying in the form of a solder. They tend to react with oxygen and with certain other materials at temperatures approaching their melting points.
Sn	231.89	
Zn	419.5	
Sb	630.5	
Al	659.7	Aluminum has been used in the semiconductor industry to form low-temperature electrical contacts. It can be applied by vacuum deposition techniques. Thermocompression or ultrasonic bonding techniques can be used to bond gold wires to aluminum.
Ag	960.5	Silver is relatively inert with respect to gas atmospheres that would normally surround a ceramic specimen, but it will react with certain ceramic samples. Silver contacts can be applied in the form of paints or by vacuum deposition techniques.
Au	1063	Gold is inert to a wide range of materials and gas atmospheres. Gold can be applied to ceramic specimens via vacuum deposition or in the form of paints and paste. Two forms of paint are available, Engelhard No. 5364, which is a cold-setting paint, and Engelhard No. 4813, which is a furnace-cured paint. The paste, Engelhard No. 601-FM, is a furnace-cured paste.[a]
Cu	1083	Copper electrodes can be applied by vacuum deposition. This material will react in a

EXPERIMENTAL TECHNIQUES

		deleterious fashion with gas atmospheres that are not inert.
Ni	1455	Nickel electrodes can be applied by electrodeless plating techniques. Contacts of this material would tend to be reactive at elevated temperatures. At low temperatures this contact material should not be used in a magnetic field, due to its large permeability.
Pd	1549.4	Palladium metal is useful as an electrical contact up to about 1200°C, and may readily be applied by vacuum deposition methods.
Pt	1773.5	Platinum is a widely used inert material for high-temperature electrical contacts and is used to approximately 1550°C. Platinum has been found to deteriorate in highly reducing atmospheres at temperatures above 1000°C. Contacts may be applied by sputtering techniques or by using furnace-cured paints or pastes. Vacuum deposition techniques using resistance heating have not been found to be successful with this metal because of its tendency to alloy. Engelhard No. 05 platinum paint and Engelhard No. 6926 platinum paste are unfluxed, while Engelhard No. 6082 platinum paste is fluxed.[a]
Rh	1966	Rhodium is seldom used as a contact material directly, but is often used as an alloying agent with platinum to form a higher melting alloy.
Ir	2454	Iridium metal is brittle but quite inert and can be used as an electrode at very high temperatures.
C (Graphite)	3500 (sublimes)	Graphite electrodes can be applied to ceramics in the form of colloidal suspensions called dag. This material can withstand high temperature but must be kept in a reducing atmosphere.

---
[a] Engelhard Industries, East Newark, New Jersey.

phase transformations that might occur and the relative thermal expansion coefficients of the materials being joined.

With the above considerations in mind, the various methods available to place electrodes on ceramic materials can be explored. The first aspect that would require attention would be the choice of an appropriate electrode material for the particular ceramic sample, temperature range and ambient atmosphere to be investigated. Higher-temperature studies usually dictate the use of the metals such as Ag, Au, Pd, Pt, Pt-Rh, Ir, etc. This is due to melting-point and reactivity considerations.

The electroding problem then narrows to one of forming a reliable mechanical bond between the ceramic sample and its electrode, and joining lead wires to the electrode material. Several purely mechanical techniques are available to accomplish this end. Contacts may be held to the sample by mechanical forces, as shown in Fig. 4(a), or wrapped tightly around the sample, as shown in Fig. 4(b). The position of the wrapped wires is better defined if shallow grooves are cut into the surface of the specimen. Contacts may also be held to the specimen by locating wires in drilled holes or sawed slots, as shown in Fig. 5. In Fig. 5(a) the wire is forced directly into the slot and, therefore, the kerf width and wire size must be adjusted accurately to yield a tight fit. In the drilled-hole methods, as shown in Fig. 5(b) and 5(c), a bead can be formed on the wire to be implanted to yield a tight connection. The integrity of all of the above connections can be improved if the surface, slots, grooves or holes are first coated with a thin layer of the metal composing the wire, or if a metal coating is placed over the configuration after the connection is made. These coatings may be applied by using such methods and materials as vacuum deposition, sputtering, flame spraying, electroless processes, soldering, cold-setting paints, and furnace-cured paints and paste. The method chosen would depend on the electrode materials being used. (See Table 3)

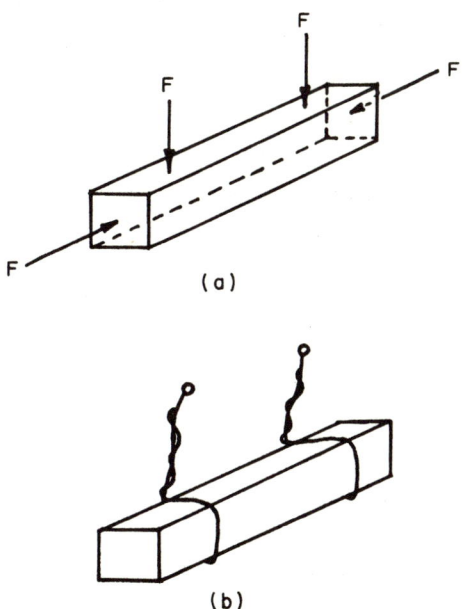

FIG. 4. (a) Sample having pressure contacts which are either weighted or spring loaded. (b) Specimen with wrapped wire contacts.

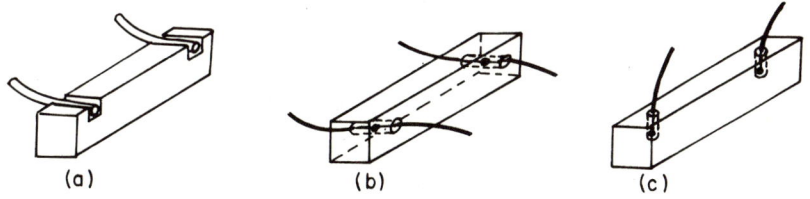

FIG. 5. (a) Specimen with wire-slot contacts. (b,c) Specimens hsving drilled hole contacts.

A second approach to achieve the same end would be first to coat the area that is to define the contact with a metal and then to bond a lead wire to this contact. The various aforementioned methods may be used to coat the contact area with a metal. At this point various bonding techniques that can be used to form a metal-metal junction must be explored.

A straightforward means of attaching a wire to the contact area would be by welding. Spot-welding, thermocompression, and ultrasonic welding methods would be appropriate for this purpose. Recently laser-welding techniques have been developed which would fill this objective. It is found, however, that the specialized welding equipment required is often not readily available. Two methods that have been particularly useful when working with platinum wires and contacts are shown in Fig. 6. In Fig. 6(a) a platinum coating is first laid down using Engelhard 6082 platinum paste. The platinum wire, which is held in position, is coated with additional platinum paste and fired. The same idea is presented in Fig. 6(b), except that a platinum foil is attached to the initial layer of platinum using furnace-cured paste and the wire is spot-welded to the foil. The method depicted in Fig. 6(c) takes advantage of the phase diagram of Pt-Au, shown in Fig. 7. An initial platinum coating is fired on the contact area using Engelhard 6082 platinum paste. A mixture of gold and platinum powder is sprinkled on this surface and a beaded platinum wire is held in position using a jig designed for this purpose. This configuration is then heated to approximately 1200°C. The gold melts, forming a liquid phase, and alloying occurs. It is observed that the gold melts at 1063°C. As alloying occurs the platinum content and melting point of the alloy increase. When the melting point reaches 1200°C no further liquid phase is present, but solid-state diffusion continues. Therefore by adjusting the amount of gold used and the temperature and time of heat treatment, any specified operating temperature for the electrode may be achieved.

Slip-cast and cementing techniques have also proven useful. These are shown in Fig. 8. In Fig. 8(a), alumina cement is mixed with platinum powder, yielding a conductive cement. This is then used to cement the platinum wire either to the ceramic specimen or

EXPERIMENTAL TECHNIQUES

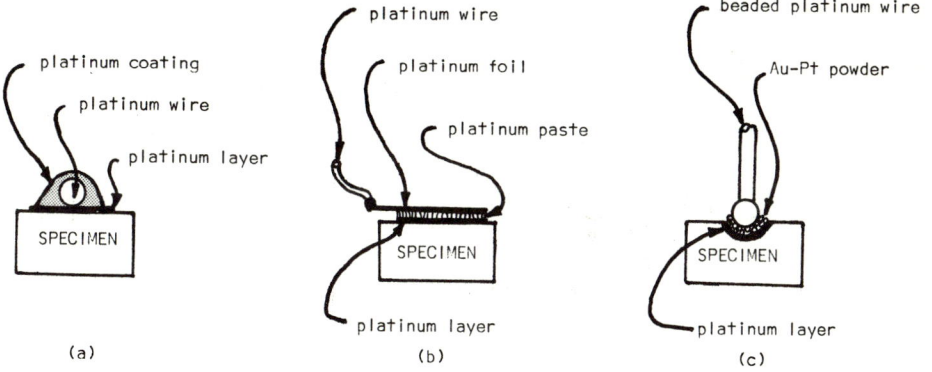

FIG. 6. Three configurations used to affix a platinum wire to a platinum contact which had previously been applied to the sample.

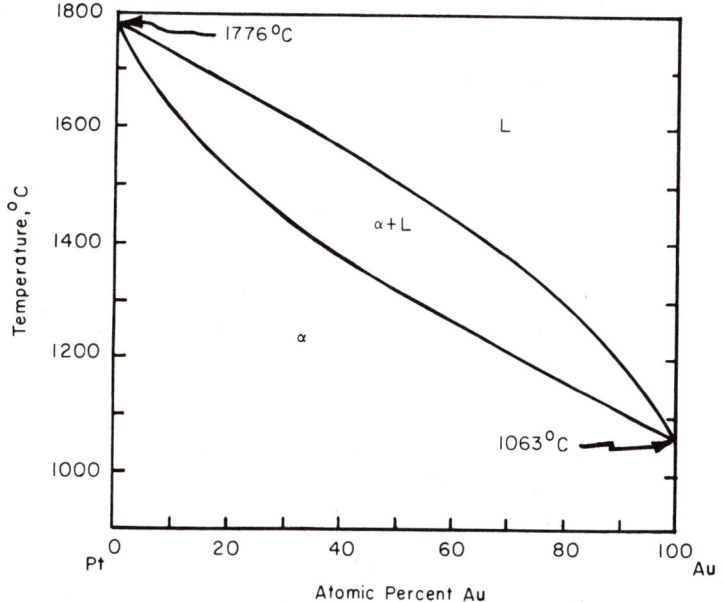

FIG. 7. Binary phase diagram for Pt-Au.

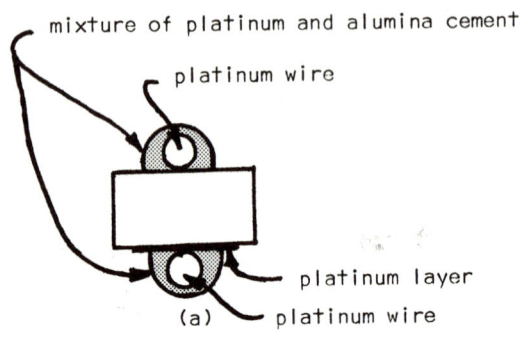

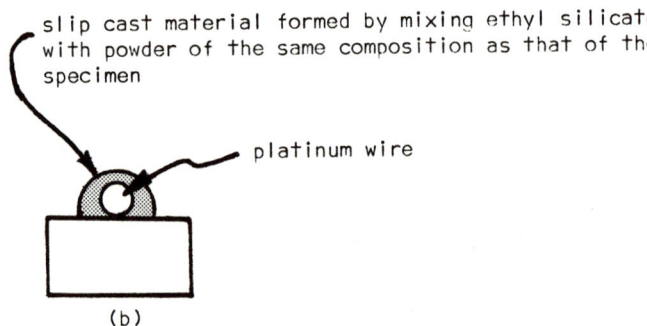

FIG. 8. (a) Two configurations by which a platinum wire can be connected to a ceramic specimen using alumina cement. (b) Platinum contact held to a specimen using a slip-cast material formed from a powder of the same composition as that of the specimen.

to a platinum coating on the ceramic specimen. The latter is more desirable when reaction occurs between the alumina cement and the ceramic substrate. In Fig. 8(b), instead of using a cement a slip-cast material is formed by mixing ethyl silicate with powder material of the same composition as the ceramic specimen. This is used to bond the platinum wire directly to the ceramic specimen. The silicon content in the final slip-cast material has been found to be less than 0.1%, and does not appreciably affect measurements made at elevated temperature or measurements made on highly doped ceramic materials. A slight dimensional error is introduced, since the

EXPERIMENTAL TECHNIQUES   57

slip-cast material acts as an extension of the sample. Contacts
made by the slip-cast method have been found to be very strong and
easy to apply.

C. SELECTION OF MATERIALS FOR CONSTRUCTING MEASURING SYSTEMS

The measuring systems discussed below in Sections III-VII
usually include an experimental apparatus holder, a heating or furnace system, and a temperature sensing device. The materials that
are used for each of these functions may be selected on the basis
of a comparison between the properties of a material (melting point,
thermal conductivity, strength at elevated temperatures, thermal
coefficient of expansion, resistance to thermal shock, resistivity,
chemical inertness to the ambient surrounding, etc.) and the environment (maximum temperature, heating rate, atmosphere, etc.) to
which the system is subjected.

For purposes of convenience the materials used for the above
functions will be described in terms of three categories: ceramics,
materials for resistance heaters, and thermocouples. The discussion
is limited primarily to a brief description of the most frequently
used materials in these categories.

1. Ceramics

A general presentation of the properties and behavior of typical
ceramic materials used in the construction of experimental systems is
given by Seybolt and Burke [14]. At temperatures below 1100°C quartz
is a frequently used material because of the following advantages:
it is gas tight; it has good mechanical strength, chemical inertness,
low coefficient of expansion, and high resistivity; and it is transparent. Because of its low coefficient of expansion the thermal
shock resistance of quartz is excellent.

Mullite may be used up to 1500-1600°C. The thermal shock resistance however is much poorer than quartz. It does have the advantage that it may be sealed directly to pyrex glass. High-purity alumina may be used up to 1900°C provided it is heated and cooled slowly. Relative to other ceramics it possesses high mechanical strength, low permeability, a high volume resistivity, low dielectric losses, and good thermal shock resistance. It is probably one of the most important and useful materials for constructing experimental apparatus and furnace tubes subjected to high temperatures.

For temperatures above 1800-1900°C refractory ceramics such as magnesia, stabilized zirconia, and thoria may be used. Magnesia has a high melting point, but it has the disadvantage that it has a high vapor pressure and starts to volatilize above 1900°C in an oxidizing atmosphere and under vacuum at about 1500°C. Although stabilized zirconia and thoria may be used at very high temperatures (up to about 2500°C), particular care must be exercised to avoid breakage from thermal shock. Another disadvantage with stabilized zirconia is that it becomes a fairly good electrical conductor at elevated temperatures and is thus not suitable as an electrical insulator.

The most frequently used materials for constructing an experimental apparatus are quartz and alumina. Specific information about the availability of the various shapes and sizes of these materials and possible applications may be obtained from the commercial producers of these products.

2. Materials for Resistance Heaters

Although high temperatures may be obtained using different heat sources [14] (e.g., induction, gas heating, resistance), because of its ease of control the most frequently used method for the measuring systems described in Sections III-VII is resistance heating.

For furnace temperatures up to 1100°C one of the most popular electrical resistance elements is an 80:20 nickel-chromium alloy. An alloy of iron, chromium, and aluminum (Kanthal) may be used up to about 1300°C. Platinum wire is frequently used as a furnace winding for use up to 1500°C. The addition of rhodium to platinum increases the maximum useful temperature up to 1800°C. Because of evaporation losses, usually as an oxide of platinum, the lifetime of the furnace decreases rapidly with increasing temperature. Nonmetallic heating elements made from silicon carbide and $MoSi_2$ may also be used up to 1500°C and 1700°C, respectively, in oxidizing atmospheres.

Mo and W wire or ribbon, wound on an alumina tube, may be used up to 1800-1900°C under vacuum or in a protective inert-gas atmosphere or a reducing atmosphere containing hydrogen. Tantalum may also be used in the same protective atmosphere, but because of its affinity for oxygen should not be placed in contact with an oxide support tube.

3. Thermocouples

Temperatures may be measured using several different techniques [14] (e.g., thermocouple, resistance thermometry, radiation thermometry). Since temperature measurements in the systems described in Sections III-VII are usually obtained with a thermocouple, this section is limited primarily to a brief discussion of commonly used thermocouples.

Copper-constantan thermocouples may be used for temperatures less than 0°C and up to 500°C in mildly oxidizing or reducing atmospheres. Constantan is recommended for low temperatures because the emf output is high and errors due to inhomogeneities are small because the homogeneity of the component wires can be maintained better than with other base-metal wires.

For temperatures up to 700-800°C thermocouples of iron-constantan may be used in reducing atmospheres, and chromel-constantan couples may be used under vacuum, or in an inert, mildly oxidizing, or reducing atmosphere. Chromel-alumel thermocouples may be used to 1100°C continuously without serious deterioration, and intermittently to about 1300°C. Protection from reducing atmospheres should be used since the emf of the couple will change with time in this type of environment.

Although there are a number of different Pt-Rh alloys used in thermocouples, the three types most frequently used are Pt vs Pt-10%Rh, Pt vs Pt-13%Rh, and Pt-6%Rh vs Pt-30%Rh. They have a high resistance to oxidation and corrosion. However, hydrogen, carbon, and metal vapors can contaminate a Pt-Rh thermocouple. The maximum temperature for continuous use of the first two of these Pt-Rh couples is 1500°C and for the Pt-6%Rh vs Pt-30%Rh couple it is 1700°C.

Three types of tungsten-rhenium thermocouples, W vs W-26%Re, W-5%Re vs W-26%Re, and W-3%Re vs W-25%Re are in common use for measuring temperatures up to 2400°C continuously with proper insulating materials and intermittently as a bare wire up to 2800°C. These alloys have inherently poor oxidation resistance and should be used under vacuum, in hydrogen or in inert atmospheres.

III. THERMODYNAMIC MEASUREMENTS

Thermodynamic studies in conjunction with theoretical defect-structure analysis can be used to provide a better understanding of electrical conduction in nonstoichiometric ceramic phases. These thermodynamic studies are generally concerned with determining

(a) the phase boundaries of a single-phase region as a function of the state variables;

(b) the thermodynamic properties of a single-phase region as a function of the state variables.

As an example, consider a nonstoichiometric binary compound, MX_2-x in equilibrium with its vapor, which has pressure P_{X_2}; then according to the phase rule $C = 2$, $P = 2$, and $F = 2$. In this case the composition x is a function of the variables T and P_{X_2} (i.e., $x = f(T_1 P_{X_2})$. If a second phase is present, $P = 3$ and $F = 1$ and the

EXPERIMENTAL TECHNIQUES

value of x at the phase boundary will be uniquely defined by the temperature. At constant temperature, the value of P_{X_2} will vary continuously in the single-phase region, whereas in the two-phase region it will be a constant. Thus by measuring the vapor pressure as a function of composition at constant temperature the phase boundaries of the compound can be determined.

As described in Chapter 3, information about the defect structure of a nonstoichiometric compound can be obtained from the relationship between the thermodynamic properties and the state variables (e.g., temperature and composition). In general thermodynamic data of this type are obtained using either electrochemical-cell or gas-equilibrium measurements.

A. GALVANIC-CELL MEASUREMENTS

Emf measurements on galvanic cells using either liquid or solid electrolytes have been employed for many years to obtain thermodynamic data. The principal advantage of cells with solid electrolytes is that they may be used at high temperatures because the solubility of the electrodes is usually quite low in solid electrolytes. In addition the container problem associated with liquid electrolytes is eliminated. Until recently galvanic-cell measurements using solid electrolytes have been limited primarily to solid halides or porcelain- or glass-type electrolytes. However, the application of these electrolyte systems has been rather limited. The work of Kiukkola and Wagner [23] on an oxide electrolyte system based on zirconia solid solutions has done much to stimulate the recent development of high-temperature galvanic cells. The applications of galvanic cells are based primarily on oxide and fluoride electrolyte systems that are reversible to oxygen and fluorine, respectively. Thermodynamic data on ceramic oxide systems (e.g., nonstoichiometric oxides, metal/metal-oxide mixtures, oxide solid solutions, spinels, silicates, molybdates, tungstates, etc.) have been obtained using oxide electrolytes. Thermodynamic data for the formation of fluorides, carbides, borides, and phosphorides have been obtained from galvanic cells using fluoride electrolytes. Recently there have been several excellent review articles on thermodynamic measurements using solid electrolyte systems [24-29].

The primary objective of this section is to describe experimental methods for obtaining thermodynamic data on nonstoichiometric compounds. Since local thermodynamic equilibria are required for data of this type, the galvanic-cell measurements are generally made at elevated temperatures using a solid electrolyte. Although a few thermodynamic studies of nonstoichiometry have been made on compounds such as sulfides, tellurides, and fluorides, the vast majority of investigations have been made on nonstoichiometric metal oxides. Therefore the discussion of emf measurements is limited to galvanic cells using solid oxide electrolytes. A general review of electrochemical cells incorporating solid oxide electrolytes has been given by Steele [26]. Because of the emphasis in this section on phase stability and nonstoichiometry of metal oxides, the description of electrochemical cells is confined to the following types of cells:

Cell I   M, MO_x | solid oxide electrolyte | reference electrode A, AO (O_2, air, etc.)

(e.g., M = Fe [23], Cu [23], Ni [23], Co [23], Mo [30], W [31], Mn [32], etc.)

Cell II MO_x, MO_y | solid oxide electrolyte | reference electrode A, AO (O_2, air, etc.)

(e.g., M = Fe [23], Mn [33], Ti [34], Nb [35], etc.)

Cell III MO_x | solid oxide electrolyte | reference electrode A, AO (O_2, air, etc.)

(e.g., M = Ti [34,36], Nb [35,36], U [37], Ce [38], Fe [39], Ni [40], etc.)

The cell emf E for Cells I, II and III is given by the well known reaction

EXPERIMENTAL TECHNIQUES 63

$$E = \frac{RT}{4F} \ln \frac{P''_{O_2}}{P'_{O_2}} \qquad (3)$$

where P''_{O_2} and P'_{O_2} are the equilibrium partial pressure of oxygen at the cathode and anode, respectively, R is the gas constant, F is the Faraday constant, and T is the absolute temperature. The polarity of the cathode is positive and it corresponds to the electrode with the largest equilibrium partial pressure of oxygen. Thus if one knows the oxygen partial pressure of the reference electrode and measures the emf as a function of temperature the equilibrium value of oxygen pressure at the working electrode may be determined. In addition to providing useful thermodynamic data, Cells I and II can be used to determine the range of oxygen pressure over which the nonstoichiometric compound MO_x is stable at a given temperature. For example, consider the nonstoichiometric oxide wüstite, Fe_xO, which may coexist in a two-phase region with either iron or magnetite, Fe_3O_4, above approximately 560°C. The equilibrium oxygen partial pressure of wüstite coexisting with iron and magnetite may be determined from emf measurements in Cells I and II, respectively. Rizzo et al.[39], using air as a reference electrode, measured the emf of Cell I for the Fe, Fe_xO equilibria and Cell II for the Fe_xO, Fe_3O_4 equilibria. From these emf measurements they obtained the following expression

$$\log P_{O_2} = \frac{-27,295 \pm 85}{T} + 6.6115 \pm 0.0218 \qquad (4)$$

for Fe, Fe_xO equilibria and

$$\log P_{O_2} = \frac{-33,210 \pm 20}{T} + 13.354 \pm 0.064 \qquad (5)$$

for the Fe_xO, Fe_3O_4 equilibria.

Emf measurements on Cell III have been used to determine the equilibrium oxygen partial pressure of nonstoichiometric oxides as a function of composition and temperature. Emf measurements on Cell III have employed two basic techniques:

(a) Measurements are made on several nonstoichiometric oxide electrodes with different fixed compositions as a function of temperature.

(b) Isothermal measurements are made on a nonstoichiometric oxide whose composition is varied electrochemically by a coulometric titration technique.

Steele [24], in his review article on electrochemical cells with solid oxide electrolytes, has derived a generalized expression for the emf of a concentration-type cell and has shown under what conditions Eq. (3) is valid. The derivation of Eq. (3) assumes the following

(a) that local thermodynamic equilibria exist at the electrode-electrolyte interfaces (i.e., the chemical potential at the electrodes is well defined;

(b) a concentration-type cell where the two electrodes are reversible to the same ionic species in the electrolyte (i.e., for a solid oxide electrolyte the ionic species is the oxygen ion);

(c) that ionic transference t_i is virtually unity in the electrolyte;

(d) that the diffusion or junction potential is negligible; and

(e) that no intermediate compound forms as a result of the reaction at the electrode-electrolyte interfaces.

Only assumption (d) will be discussed here; the other assumptions will be discussed in the appropriate sections on selection of electrolytes and electrodes.

Steele has shown that the diffusion potential is negligible

(a) if $t_i = 1$ and the dissolution of electrodes into the electrolyte is negligible;

(b) if dissolution of the electrodes into the electrolyte occurs, and the transference number of the ionic species to which the two electrodes are reversible is unity (i.e., for solid oxide electrolyte $t_{O^{2-}} = 1$).

EXPERIMENTAL TECHNIQUES

He has also shown that the appropriate expression for the emf of the cell is

$$E = \frac{RT}{4F} \int_{P'_{O_2}}^{P''_{O_2}} t_{ion} \, d \ln P_{O_2} \qquad (6)$$

if

(a) there is negligible dissolution of the electrode components into the electrolyte and the ionic transference number is less than 1;

(b) dissolution of the electrode components into the electrolyte occurs provided the sum of the transference number $t_{O^{2-}}$ of the ionic species to which the two electrodes are reversible and the electronic transference number t_e is equal to unity (i.e., for solid oxide electrolytes $t_{O^{2-}} + t_e = 1$).

1. Selection of Electrolyte Material

On the basis of the above discussion regarding the assumptions used in deriving the expression for the emf according to Eq. (3), a good solid electrolyte should have the following characteristics:

(a) the ionic transference number t_i should be essentially unity if no dissolution of the electrodes into the electrolyte occurs;

(b) if dissolution of the electrode into the electrolyte occurs, the transference number of the ionic species to which the two electrodes are reversible should be essentially unity.

Another important assumption made in the derivation of Eqs. (3) and (6) was that local thermodynamic equilibria (i.e., well-defined chemical potentials or partial pressure of oxygen) existed at the electrode-electrolyte interfaces. This assumption may be invalidated because of either internal or external current leakage paths. The internal "short circuiting" occurs because of electronic conductivity in the electrolyte. The result of this effect, for example, in an oxide electrolyte such as $Zr_{0.85}Ca_{0.15}O_{1.85}$ is the transfer of oxygen from the cathode through the electrolyte to the anode. If the electrochemical reaction at one of the electrodes is rate controlling, this flux of oxygen will produce an undefined higher oxygen partial pressure or chemical potential at this electrode and the emf of the cell will be decreased.

On the other hand, if the electrochemical reactions at the electrodes are not rate controlling, then for cells of type III the flux of oxygen may significantly change the composition of the non-stoichiometric oxide electrode. It should be noted that for electrolytes with a high total conductivity, the electronic currents can be significant even when the electronic transference number is small ($\sim 10^{-3}$).

At elevated temperatures external current leakage paths may arise because of conduction through the gas phase. This problem can generally be eliminated by using a cell assembly with a guarded electrode arrangement. The solution to the problem will be discussed in the section on cell assemblies.

Etsell and Flengas have made a comprehensive review of the electrical properties of solid oxide electrolytes [41]. Most of the oxide electrolytes that have been used to obtain thermodynamic data are based on solutions of ZrO_2 or ThO_2 doped with either CaO or Y_2O_3. The ionic conductivity in these materials is primarily controlled by the oxygen-ion conductivity because the cationic contribution is several orders of magnitude smaller. For example, at 1000°C and oxygen pressure of 1 atm the ionic conductivity of each ion in the frequently used electrolyte $ZrO_{0.85}Ca_{0.15}O_{1.85}$ has been summarized by Steele [24] as follows:

Total electrical conductivity	4.0×10^{-2} ohm^{-1} cm^{-1}
Oxygen ion conductivity	4.0×10^{-2} ohm^{-1} cm^{-1}
Zirconium ion conductivity	1.0×10^{-12} ohm^{-1} cm^{-1}
Calcium ion conductivity	1.1×10^{-13} ohm^{-1} cm^{-1}

The values of the ion conductivity were calculated from diffusion data using the Nernst-Einstein relationship. These and other data indicate that the oxygen-ion transference number for this electrolyte is essentially unity. Thus the problem of a junction potential arising from the dissolution of the electrode into the electrolyte is minimized by using this electrolyte.

Although the contribution of each ion to the total conductivity may be calculated, the electronic conductivity has not been accurately determined. The results of several investigators employing different methods of obtaining electronic transport (e.g., emf technique and polarization measurements) were reported by Etsell and Flengas [41]. Although there is a great deal of scatter in the reported values of the critical oxygen pressures corresponding to the ionic transference number t_i at 1000°C (e.g., t_i = 0.99 for log P_{O_2} = -16 to -23, and t_i = 0.50 for log P_{O_2} = -24 to -31) it is generally accepted that t_e increases with decreasing oxygen pressure. Some of the scatter in these data may be attributed to the use of different experimental methods for obtaining the transference number. The effect of impurities on electronic conductivity, however, may provide a more reasonable explanation for scatter in these results. For example, commercially prepared electrolytes generally have a higher electronic conductivity than those prepared in the laboratory [26].

As discussed above, the principal limitation in the use of zirconia-calcia electrolytes is the presence of a small electronic conductivity, particularly at low oxygen pressures. Electrochemical-cell measurements have also been made utilizing thoria-based electrolytes. Etsell and Flengas [41] have reviewed the literature on the electrical properties and thermodynamic measurements of this electrolyte. Although the thoria-based electrolytes exhibit some p-type conductivity [42] at high oxygen pressures, the principal advantage is that they exhibit less electronic conductivity at low oxygen pressures [43]. Thus emf measurements may be made at lower oxygen pressures with ThO_2-based electrolytes than with ZrO_2-based electrolytes. It should also be noted that the total conductivity of the ThO_2-based electrolytes is lower than that of the ZrO_2-based electrolytes. Thus the open-circuit oxygen transport between the electrodes will be reduced and there will be less tendency to polarize the electrodes.

Errors in emf measurements because of polarization of electrodes may also occur from permeability of oxygen through a porous electrolyte. Thus care should be exercised in the preparation of solid electrolytes, to avoid the formation of open pores. The techniques for preparation of zirconia- and thoria-based electrolytes will not be discussed here since they have been reviewed by Etsell and Flengas [41].

2. Selection of Electrodes

In the derivation of Eq. (3) it is assumed that a well-defined chemical potential of oxygen or equilibrium oxygen partial pressure exists at each electrode-electrolyte interface. Open-circuit current-leakage paths, reactions of the gaseous environment with the electrodes, or reactions between the electrodes and the electrolyte may disturb local thermodynamic equilibrium and produce an undefined chemical potential of oxygen at the electrode-electrolyte interfaces.

Current leakage between the electrodes may be caused by either internal or external paths. Internal current paths arise because of electronic conduction in the electrolyte, whereas external current paths are caused by surface conduction or conduction through the gas phase. The net result of current leakage is to produce a flux of oxygen from the cathode to the anode. If the chemical reactions and diffusion at the electrodes are not sufficiently rapid to maintain thermodynamic equilibrium an undefined oxygen chemical potential will result and the emf of the cell will be decreased. If a steady-state condition is not obtained the emf of the cell will continue to decrease with time.

Internal "short circuiting" may occur even when $t_e < 0.01$, because the electronic conductivity can be significant if the total electrical conductivity of the electrolyte is high. For example,

EXPERIMENTAL TECHNIQUES

Steele [26] has given the following calculation for a $Zr_{0.85}Ca_{0.15}O_{1.85}$ pellet of 1 cm^2 cross section and 1 mm thickness with a resistance of 2.5 ohms at $1000°C$. Assuming the open-circuit emf is 500 mV an electronic transference number of 10^{-2} would allow a current of 2 mA to flow provided the electrochemical reactions at the electrodes are not rate-controlling. The internal current leakage may be reduced by selecting an electrolyte with a smaller total conductivity (e.g., a thoria-based electrolyte instead of a zirconia-based electrolyte). External current-leakage paths may be minimized by selecting the appropriate cell assembly. This problem is discussed in more detail in Section III,A,3.

Reactions between the electrodes and the electrolyte, and reactions of the gaseous environment with the electrodes and open-circuit leakage paths may disturb local thermodynamic equilibrium and produce an undefined chemical potential of oxygen at the electrode-electrolyte interfaces. The first consideration in selecting an electrode should be its compatibility with the electrolyte (i.e., the mutual solubility of the electrode and electrolyte should be negligible and no intermediate reaction products should form between the electrode and electrolyte). It should not be difficult to select a reference electrode which meets this requirement since a large number of electrode systems have been shown to be compatible with the aforementioned solid oxide electrolyte systems. Since the nature of the investigation determines the electrodes on which thermodynamic measurements are to be obtained, the electrolyte should be selected to minimize any reaction with this electrode.

Electrochemical-cell measurements are generally made in an inert atmosphere or under vacuum. Gaseous impurities such as O_2, CO, CO_2, H_2, H_2O etc., are, however, generally present in these atmospheres. Therefore an electrode will "getter" oxygen if the oxygen chemical potential of the electrode is less than that of the surrounding atmosphere and will furnish oxygen to the atmosphere if the oxygen chemical potential of the electrode is greater. To minimize this problem the inert atmosphere is generally purified before introducing the gas into the cell assembly [35]. The surrounding atmosphere may also act as a carrier gas for the transfer of oxygen between the electrodes. This problem is generally eliminated by using a cell assembly with separate electrode chambers. This type of cell assembly will be discussed in Section III,A,3.

Although limited information about the electrochemical kinetics of cells with solid oxide electrolytes is available, the polarization behavior of the following reference electrodes has been discussed by Steele [26]. Porous platinum-oxygen electrodes and iron-wüstite electrodes exhibit insignificant polarization at 1000°C with current densities as high as 100 $\mu A/cm^2$. The diffusion coefficient for iron in wüstite is a function of the nonstoichiometry and varies between 10^{-5} and 10^{-7} cm^2/sec at 1000°C. It is interesting to note that cation diffusion coefficients in nickel oxide are 10^{-10}-10^{-11} cm^2/sec at 1000°C and Ni-NiO electrodes are observed to exhibit significant polarization with current densities less than 5 $\mu A/cm^2$ at 1000°C.

Information about polarization studies on nonstoichiometric electrodes is very important, particularly when coulometric titration techniques are used. For example, coulometric experiments by Alcock et al. [44] on cells with Fe-FeO and nonstoichiometric rutile electrodes indicate that, after titration of rutile electrodes with currents as low as 1-5 $\mu A/cm^2$ at 1000°C, equilibration took at least 24 h. The diffusion coefficient of oxygen in rutile is approximately 10^{-13} sec^{-1} at 1000°C.

Although the magnitude of the diffusion coefficient for an electrode appears to be an indication of the reversibility of an electrode, the reversibility of the electrodes should be determined experimentally. This is generally accomplished by drawing current in either direction and noting whether the observed voltage returns to the original value.

Another requirement of the electrodes assumed in the derivation of Eq. (3) is that the electronic transference number of the electrode should be essentially unity. Otherwise there would be a gradient in

EXPERIMENTAL TECHNIQUES 71

the electrochemical potential of the electrons across the electrode with a resulting drop in voltage across the electrode.

3. Cell Assemblies

Typical arrangements that have been used to measure the emf of Cells I, II, and III are shown in Figs. 9-11. The simple cell assembly illustrated in Fig. 9(a) was the type originally used by Kiukkola and Wagner [23] to measure the emf of Cells I and II. Some emf measurements using this cell assembly have also been made on cells of type III [34,35,37]. This type of cell assembly generally functioned satisfactorily as long as the oxygen chemical potential of the electrodes to be measured was higher than that of the reference electrode. If the chemical potential of oxygen was much lower than the reference electrode, oxygen was probably transferred from the reference electrode through the gas phase by impurities in the gas (e.g., H_2, CO) to the electrode to be measured. Schmalzreid [45] attempted to reduce this interaction between the electrodes and the gas phase by enclosing part of the electrode surface with electrolyte material, as shown in Fig. 9(b).

The most satisfactory method of eliminating the transfer of oxygen between the electrodes via the gas phase is to use a cell assembly with two separate electrode compartments. The cell assemblies shown in Fig. 10 differ from the cell assemblies shown in Fig. 9 in that a stabilized zirconia electrolyte in the form of a closed-end tube is generally used instead of a pellet-shaped electrolyte. When emf measurements are made on an electrode having a very low chemical potential of oxygen (i.e., $P(O_2) \leq 10^{-20}$ atm at 1000°C), the electronic contribution in the stabilized zirconia electrolyte may become significant. To avoid this problem, Markin and Bones [46] separated the closed zirconia tube from the electrode having the low oxygen potential by means of a pellet of stabilized thoria, as shown in Fig. 10(b). The reason for this modification is that the stabilized thoria electrolytes have lower electronic transference numbers and electronic conductivities than stabilized

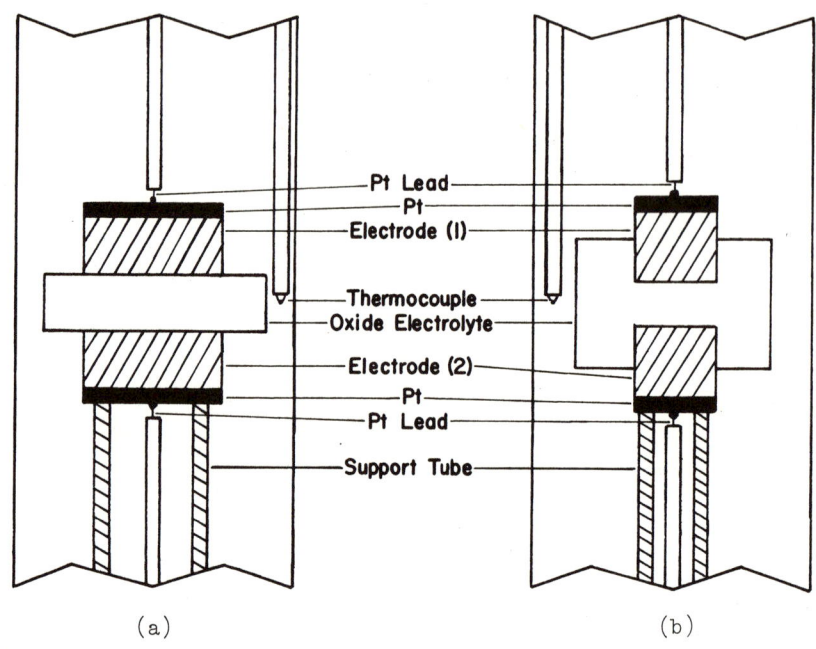

FIG. 9. Schematic of simple cell assemblies used to measure the emf of Cells I-III

zirconia electrolytes at low oxygen potentials. When coulometric measurements are made, the electrodes may become polarized after coulometric titration. To avoid polarization of the reference electrode, both a working electrode and a reference electrode may be used, as shown in Fig. 10(c). The advantage of this cell assembly is that the working electrode may act as source or sink for oxygen during the coulometric titration and the reference electrode will not be polarized, since it is used only for the open-circuit emf measurements.

Because the nonstoichiometric electrode is usually polarized after a coulometric titration, the emf of the cell is generally measured as a function of time to determine when equilibrium is obtained. This equilibration process is generally diffusion-controlled and the time required for equilibration depends upon a number of factors, e.g., the rate and amount of oxygen titrated,

EXPERIMENTAL TECHNIQUES

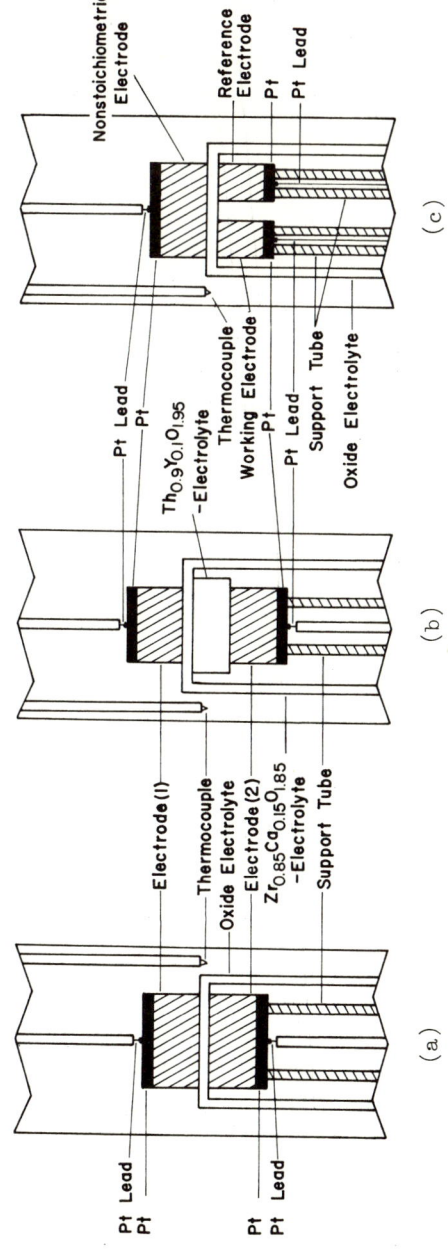

FIG. 10. Schematic of cell assemblies with closed end electrolyte tubes used to measure the emf of Cells I-III.

the diffusion coefficient of the ions, the thickness of the electrode, the given size and porosity of the electrode. Alcock et al. [47] in their coulometric study of nonstoichiometric rutile (TiO_{2-x}) found that with titration currents of 1-5 $\mu A/cm^2$ equilibration took at least 24 hours at 1000°C. One of the major difficulties with the cell assemblies shown in Figs. 9 and 10 is that the electrodes may react with impurities in the ambient atmosphere. The electrodes may be oxidized or reduced depending on whether the chemical potential of oxygen in the atmosphere is less than or greater than the oxygen potential of the electrode. This problem becomes more severe when a coulometric titration technique is used since longer equilibration times are required. As discussed above, if an inert gas is used as an ambient atmosphere it is purified to reduce the impurities prior to admitting the gas into the cell assembly [35]. In a further attempt to minimize this problem, some investigators have also used a "gettering" material (e.g., the same material as the electrode to be measured) in the vicinity of the electrode so that the chemical potential of oxygen in the gas would be close to the value of the electrode.

More recently the cell assemblies shown in Fig. 11 were developed for solid-state coulometric titrations in an attempt to overcome these problems. The cell assemblies shown in Figs. 11(a) and 11(b) have been described by Steele [26] and Tretyakov and Rapp [40], respectively. These cell assemblies are similar in that they attempt to provide the powdered nonstoichiometric phase with its own gaseous environment in an impermeable alumina crucible. The chemical potential of oxygen in this environment is controlled by coulometric titration using porous platinum-oxygen electrodes. The change in oxygen content of the nonstoichiometric phase is calculated by subtracting the change in oxygen content of the gas in the enclosure from the amount of oxygen titrated into or out of the enclosure [40].

The basic difference between these cell assemblies is in the methods used to form a seal between the closed alumina crucible and the electrolyte. Steele used liquid sodium silicate made from

EXPERIMENTAL TECHNIQUES

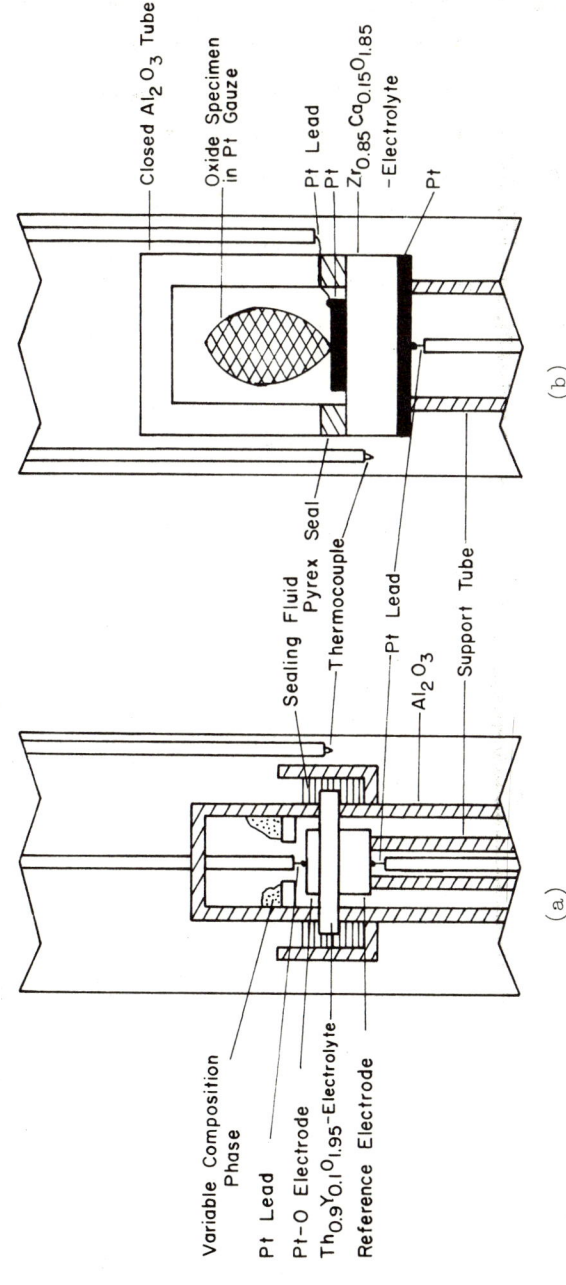

FIG. 11. Schematic of cell assemblies for solid-state coulometric titrations.

25 mole % Na_2O and 75 mole % SiO_2 with a melting point of ~ 800°C as a sealing fluid, whereas Tretyakov and Rapp used a Pyrex seal.

The cell assembly shown in Fig. 11(a) has been used to investigate the TiO_{2-x} system. Since the rutile phase is present in the form of a powder, more surface area is exposed and the diffusion length is shorter with the result that equilibration is more rapid. According to Steele [26], titration currents up to 50 $\mu A/cm^2$ have been used with equilibration times of only 1-3 h required. At compositions approaching stoichiometry (x in TiO_{2-x} < 0.001) in the temperature range 900° to 1100°C the emf exhibits a continual drift with time, 1-5 mV/h, which indicated that the cell assembly was not sufficiently impermeable to oxygen to prevent a change in the oxygen content of the gaseous environment.

Tretyakov and Rapp [40] studied the stoichiometry ranges of $NiO_{1+\delta}$ and $LiFe_5O_{8-\delta}$ using the cell assembly shown in Fig. 11(b). These experiments were limited to oxygen activities between 1 and 10^{-4} atm O_2 in the temperature range 1100-1300°K. The Pyrex seal prevented significant interaction of the enclosed electrode with the gas phase surrounding the cell when their ratio in oxygen activity was greater than 5×10^{-4} atm. According to Rapp, (a) selection of a reference electrode with a low P_{O_2} should allow the use of this cell arrangement at low P_{O_2}, (b) the use of a viscous Pyrex ring to seal the electrode chamber should be applicable for electrodes which will neither reduce nor react with silicon in the temperature range 1100-1300°K. For measurements at higher or lower temperatures, SiO_2-base glasses of other compositions and softening temperatures should be chosen.

B. GAS-EQUILIBRATION MEASUREMENTS

Gas-equilibration measurements have been used extensively for many years to obtain thermodynamic data. Kubaschewski et al. [48] have reviewed the experimental methods used to investigate heterogeneous equilibria between condensed-phase mixtures (metal carbides,

EXPERIMENTAL TECHNIQUES   77

nitrides, oxides, sulfides, chlorides, etc.) and gas mixtures which
contain one of the components of the condensed phases (H_2-CH_4, H_2-
NH_3, H_2-H_2O, H_2-H_2S, H_2-HCl, etc.). The primary objective of this
section is to describe gas-equilibration methods for obtaining thermodynamic data on nonstoichiometric compounds. In principle each
method involves the following two requirements:

(a) fixing the composition of the nonstoichiometric compound
by equilibrating with a gaseous atmosphere which has a well-defined
chemical potential or vapor pressure for at least one of the components of the compound

(b) determination of the composition of the nonstoichiometric
compounds

These are discussed below in greater detail.

1. Controlling the Composition of a Nonstoichiometric Compound

When a gas-equilibration technique is used the composition of a
nonstoichiometric compound is controlled by equilibrating the compound with an appropriate gaseous atmosphere. The minimum number
of variables that must be specified in order to uniquely determine
the composition of the nonstoichiometric phase is given by the phase
rule. For example, for a binary nonstoichiometric compound in equilibrium with the vapor of one or both of its components there are
only two independent variables. Thus the composition in the nonstoichiometric phase will vary with both temperature and the partial pressure of one of the components. If a second condensed phase
forms, there is only one independent variable and for conditions of
constant temperature the pressure will be independent of the overall
composition of the phase mixture. Therefore by measuring the composition as a function of the partial pressure of one of the components the stability range of the nonstoichiometric binary compound
can be determined. This method is illustrated in Fig. 12 where the
stability range of a hypothetical nonstoichiometric compound MX_{1-x}
is determined from a pressure-composition isotherm. The stable
gaseous form of X is assumed to be X_2 the partial pressure and P_{X_2}

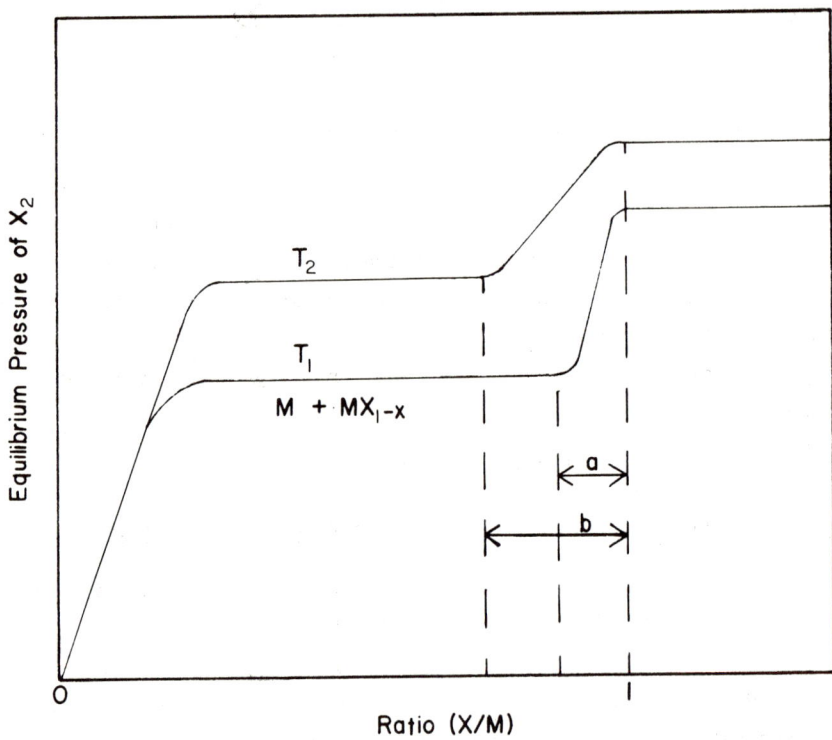

FIG. 12. Schematic diagram of the stability range of a hypothetical nonstoichiometric compound MX_{1-x} as illustrated on a pressure-composition isotherm; a = the stability range of MX_{1-x} at T_1 and b = the stability range at T_2.

is assumed to be much larger than P_M. In the isothermal regions where $M + MX_{1-x}$ coexist the partial pressure of X_2 is a constant. In the region where the nonstoichiometric compound MX_{1-x} exists, P_{X_2} varies with composition. Thus, the range of P_{X_2} over which the nonstoichiometric phase MX_{1-x} is stable at a given temperature may be determined by the intersection of the pressure-composition curve in the single-phase region with the constant pressure plateau.

The simplest method of controlling the composition of the nonstoichiometric phase is to control directly the vapor pressure of one of the components. This technique is generally applicable if

EXPERIMENTAL TECHNIQUES

only one of the components exhibits a significant vapor pressure. The major difficulty with this method occurs at low pressures (i.e., roughly when $P_i < 10^{-6}$ atm) where the gas has very little "buffering" capacity. The "buffering" capacity of a gas may be defined as the capacity of the gas to exchange component i with its environment without producing a significant change in the value of P_i. Thus, for example, consider a metal oxide in equilibrium with a gaseous atmosphere which contains only oxygen. If the value of P_{O_2} is less than approximately 10^{-6} atm a small leak in the system could alter the value of P_{O_2}. In addition, it would be difficult to change the composition of the compound by changing the value of P_{O_2} since even a small change in the stoichiometry of the sample would produce a significant change in P_{O_2}.

Either static or dynamic techniques may be employed to control directly the pressure of one of the components of the nonstoichiometric phase. In the static method the nonstoichiometric compound is usually encapsulated in an evacuated container with another condensed phase which may act as either a source or sink for one of the components of the nonstoichiometric compound. The capsule is placed in a furnace with the nonstoichiometric compound at temperature T_1 and the other condensed phase at temperature T_2. The temperature T_2 is used to control the partial pressure of one of the components in the nonstoichiometric compound. As an example of this technique consider the case where the nonstoichiometric compound $Zn_{1+x}O$ is at T_1 and the condensed phase liquid zinc is at T_2. With this arrangement zinc may be transferred through the vapor phase from the liquid to $Zn_{1+x}O$. The composition of $Zn_{1+x}O$ (i.e., the departure from stoichiometry x) is determined by controlling the values T_1 and P_{Zn}. The value of P_{Zn} is fixed by the temperature of the liquid zinc, T_2. A static technique has also been used where both the nonstoichiometric compound and another condensed phase are placed in an evacuated chamber at the same temperature. This method, which is referred to as the isopiestic technique, has been used to prepare nonstoichiometric specimens of TiO_{2-x} [34] and Nb_2O_{5-x} [35].

The pressure or partial pressure may also be controlled directly by using dynamic techniques. If the most stable state of one of the components in the nonstoichiometric compound is a gas, then the partial pressure may be controlled between approximately 1 and 10^{-6} atm by either of the two following methods:

(a) by allowing the gas to leak at a controlled rate into a vacuum system

(b) by diluting the gas with an inert gas

When method (a) is used dynamic pressures as low as 10^{-6} atm can be maintained with the aid of a variable leak and continuous pumping with a vacuum pump on the sample chamber [49]. The leak rate may be controlled using a sensitive needle valve or a small capillary. A number of different types of vacuum gauges are suitable for accurately measuring the pressure over the entire range of pressure $1-10^{-6}$ atm [50]. For example, a precision Bourdon-type differential manometer may be used for the pressure range 1 atm to 0.001 atm. A thermocouple gauge may be used for the 100-1 μ range, and for lower pressures a hot-filament ionization gauge is suitable. Care should be exercised when making absolute pressure measurements at low pressures in the presence of a temperature gradient because of pressure gradients arising from thermomolecular flow [51].

Controlled partial pressures of a gas from 1 to 10^{-6} atm can also be achieved with method (b), by blending the gas with a high-purity inert gas. Accurate mixtures may be obtained using calibrated flowmeters or by using premixed and analyzed gas-tank mixtures. If the most stable form of a component of the compound is not a gas, but a condensed phase which exhibits a relatively high vapor pressure (e.g., Zn in $Zn_{1-x}O$) then the partial pressure may be controlled by passing an inert carrier gas over a volatile condensed phase containing the component at an elevated temperature. The vapor pressure of the volatile component may then be determined from the weight loss of the condensed phase, or from the temperature of the condensed phase if the vapor pressure is known and equilibrium is obtained.

EXPERIMENTAL TECHNIQUES

This method has been used to control the vapor pressure of Zn over $Zn_{1-x}O$.

The above techniques, which control the partial pressure directly, are limited to the approximate pressure range $1-10^{-6}$ atm. The partial-pressure stability limits of the components for most nonstoichiometric compounds, however, extend over several orders of magnitude. As examples, the lower oxygen-pressure stability limit of TiO_{2-x} is 7.16×10^{-19} atm at 1000°C [34] and the range of permissible oxygen pressure for the $Fe_{1-x}O$-state phase is between approximately 10^{-11} and 3×10^{-14} atm at 1100°C [52]. Therefore when a compound is stable over a wide range of partial pressures (e.g., TiO_{2-x}) only a small portion of the homogeneous-phase field may be investigated using these techniques, and when the compound is stable only at low partial pressures (e.g., $Fe_{1-x}O$) these techniques are unsuitable for controlling the composition. To obtain lower partial pressures it is generally easier to use an indirect method to control the pressure of the more electronegative component of the nonstoichiometric compound. Thus, for oxides, nitrides, hydrides, carbides, chlorides, etc., it is usually more convenient to control the corresponding partial pressure of oxygen, nitrogen, hydrogen, carbon, chlorine, etc., in the gas phase than it is to control the partial pressure of the metal component.

As an example, the partial pressure of oxygen may be controlled at low oxygen pressures by using mixtures of CO and CO_2. For the reaction

$$CO + \tfrac{1}{2}O_2 \rightarrow CO_2 \tag{7}$$

the standard Gibbs free energy, $\Delta G°_T$, is given by Wicks and Block [53] and from equilibrium thermodynamics

$$\Delta G°_T = -RT \ln K = RT \ln \frac{P_{CO}\, P_{O_2}^{1/2}}{P_{CO_2}} \tag{8}$$

so that

$$\log P_{O_2} = 2 \log \frac{P_{CO_2}}{P_{CO}} + \frac{2 \Delta G^\circ_T}{2.303RT} \qquad (9)$$

Therefore the oxygen pressure in equilibrium with a nonstoichiometric oxide can be controlled by passing a gaseous mixture with a known ratio of CO_2 to CO over the sample at a given temperature. Using a similar thermodynamic argument it can be shown that [48]

P_{O_2} may be controlled from H_2-H_2O equilibria

P_{S_2} may be controlled from the H_2-H_2S equilibria

P_{Cl_2} may be controlled from the H_2-HCl equilibria

P_{N_2} may be controlled from the H_2-NH_3 equilibria

Controlled partial pressures using the above types of gas equilibria with two or more components in the gas phase may be obtained using either a static or a dynamic technique.

In the static technique the nonstoichiometric compound is generally equilibrated with a gaseous mixture in a closed system. The composition of the gas phase may either remain constant or be allowed to vary during the equilibration process, depending upon the technique employed. For example, if a mixture of CO and CO_2 is encapsulated in the closed system with a nonstoichiometric metal oxide, oxygen will be exchanged between the oxide and the gas phase during the equilibration process. The composition of the gas phase will vary with time until equilibrium is obtained. If this technique is used the composition of the gas phase is usually analyzed after the specimen is equilibrated with the gas phase. However, in order to maintain a fixed composition of the gas phase during equilibration, a source or sink for the equilibration species must be included in the enclosed system. Markin et al. [54] have used this type of technique of gas equilibration in a sealed silica system to obtain thermodynamic data on substoichiometric UO_2 above 1800°C. The UO_2 was equilibrated with a reference metal/metal-oxide mixture held at a lower temperature to fix the equilibrium H_2O/H_2 ratio and hence the partial pressure of oxygen at some suitable value. Oxygen was

EXPERIMENTAL TECHNIQUES 83

transferred from the UO_2 to the reference metal (Cr and Nb were convenient metals for this range of $\Delta \bar{G}_{O_2}$) via the gas phase.

When the dynamic technique is employed the nonstoichiometric compound is generally equilibrated with a constant-composition gas mixture in an open-flow system. The composition of the gas mixture may be determined by blending gases together using calibrated flowmeters or by analyzing the gas mixture.

Another method that has been used is to equilibrate a gas mixture with a condensed phase or phase mixture. One advantage of this technique is that it may be possible to obtain gas mixtures with a wider range of compositions than can be obtained by the above gas-mixing technique. Bevan and Kordis [55], for example, passed CO_2 over graphite at temperatures between 400°C and 700°C to obtain values of CO/CO_2 less than 10^{-3}. They also passed H_2 over a $Cr-Cr_2O_3$ mixture in the temperature range 900-1200°C to obtain ratios of H_2O/H_2 less than 10^{-3}. This method assumes that complete equilibrium is obtained between the gas mixture and the condensed phase or phases. With this technique the flow rate should be varied to insure that equilibrium is achieved.

The amount of impurities present in the gas may limit the range of composition that may be measured by either of these methods. For example, for $CO-CO_2$ gas mixtures, the calculation of P_{O_2} using Eq. (9) assumes that the only gases present are CO and CO_2. Thus the concentration of impurities (e.g., CH_4, H_2, O_2) which are normally present in these gases should be at least an order of magnitude less than the concentration of either CO or CO_2 to avoid error in the calculated value of P_{O_2}. The thermal decomposition of CO_2 places an upper limit on the oxygen partial pressure that can be obtained with this type of gas mixture, while a lower limit is placed by the thermal decomposition of CO. These limits were calculated by Vest [56] and are given below for a total pressure of 0.1 atm.

Temperature (°C)	P_{O_2} max (atm)	P_{O_2} min (atm)
1000	6×10^{-6}	2.3×10^{-20}
1600	1.7×10^{-3}	3×10^{-18}

The accuracy of the thermodynamic data for the gas-equilibrium reaction may also limit the accuracy of the calculated partial pressures. In addition, the system should always be checked to see that it is gas tight to avoid error from impurities arising from gas leaks. Since most gas equilibration measurements are made in a ceramic system at elevated temperatures the possibility of errors arising from permeability of gases through the ceramic container should also be considered.

For the gas-equilibration techniques which use gas mixtures another possible source of error is thermal diffusion. In the presence of a temperature gradient the gaseous specimen will tend to separate and produce a concentration gradient. This problem is more likely to occur when a static technique is used. When a dynamic technique is used this problem may be avoided by using a flow rate high enough to prevent separation of the gas mixture.

Darken and Gurry [57] investigated the effect of flow rate of CO-CO_2 gas mixtures on the iron, wüstite, CO-CO_2 equilibria and found that linear flow rates greater than 0.9 cm/sec were required to prevent separation of the CO-CO_2 gas mixtures.

2. Determination of the Composition of Nonstoichiometric Compounds

The gas-equilibration techniques used to control the composition of nonstoichiometric compounds are generally employed at elevated temperatures to insure equilibrium conditions. After equilibration the composition of the compound or the extent to which the compound departs from stoichiometry may be determined by using measuring techniques at high temperatures on equilibrated specimens or at low temperatures on quenched specimens. Particular care should be exercised in preparing quenched specimens, in order to preserve the composition characteristic of the high-temperature equilibrated state. To minimize this problem, nonstoichiometric specimens are usually cooled very rapidly in a purified inert atmosphere to minimize any reactions between the compound and its gaseous environment.

For purposes of convenience the type of techniques for measuring the composition of nonstoichiometric compounds is described below in terms of low-temperature and high-temperature techniques.

a. <u>Experimental Methods for Determining the Composition of Quenched Specimens of Nonstoichiometric Compounds</u>. The most direct method of determining the composition of a compound is by chemical analysis. Since the accuracy of a standard chemical analysis is approximately about 1 part in 10^3 the departure from stoichiometry x (e.g., x in $MX_{1\pm x}$) can only be determined with a precision of the order $\pm 10^{-3}$. The departure from stoichiometry in most compounds, however, is considerably smaller than 10^{-3}. Thus in order to detect values of x less than 10^{-3} special chemical techniques must be used.

Libowitz [58] has discussed some examples of special chemical techniques that will determine only the excess quantity of material, or some other manifestation of the excess quantity. For example, excess metal in KCl (Cl vacancies) or excess Ba in BaO (as oxygen vacancies) has been determined [59,60] by dissolving the compound in water and measuring the amount of hydrogen evolved according to the reaction

$$K + H_2O \rightarrow K^+ + OH^- + \tfrac{1}{2}H_2$$

Deviations in stoichiometry of approximately one part in 10^5 have been obtained by this method.

For many nonstoichiometric compounds, however, there are no suitable analytical chemical techniques for directly determining the deviation from stoichiometry. In these cases information about the relative change in stoichiometric may be obtained using physical methods. The physical measuring techniques may be classified into two categories: direct measurements of changes in stoichiometry, and indirect determination of changes in stoichiometry using property measurements and defect models.

A direct measurement of the change in composition of a nonstoichiometric compound relative to its composition in a reference state may be obtained using gravimetric or volumetric techniques. An example of the volumetric technique is the preparation of metal hydrides of closely known stoichiometry by adding accurately measured volumes of hydrogen gas to pure metals at precisely determined temperatures and pressures.

The simplest gravimetric method is to measure the weight change of a quenched specimen at room temperature. For example, the departure from stoichiometry (x) for Nb_2O_{5-x} was determined [35] from the per cent weight change of the specimen and the expression

$$x = \left[\frac{\% \text{ wt loss}}{100}\right]\left[\frac{\text{mol wt } Nb_2O_5}{\text{at. wt oxygen}}\right] \qquad (10)$$

The composition of specimens fired in air was assumed close enough to stoichiometry to be considered stoichiometric. All weight changes which occurred on reduction were attributed to a change in oxygen content of the specimen, because experimentally a reduced specimen, when oxidized in air at 500-1300°C, would return to its initial weight, within experimental error.

Relative changes in stoichiometry may also be inferred from measurements of properties which are sensitive to composition. For example, electrical conductivity is usually very dependent on deviations from stoichiometry. If $\sigma = ne\mu$ and μ is known from Hall mobility measurements, then n, the concentration of carriers, may be calculated from the conductivity measurements. In addition, if a simple defect model is assumed to exist which relates n to the departure from stoichiometry then the composition of a compound may be determined. The basic problem with this approach is that a simple defect model must be assumed. In addition, the mobility is

EXPERIMENTAL TECHNIQUES

usually a function of defect concentration and the presence of impurities may influence the value of n.

Other types of physical measurement which depend on deviation from stoichiometry, e.g., lattice-parameter, magnetic, and optical measurements, may also be used to determine the composition [58].

b. <u>Experimental Methods for Determining Deviations from Stoichiometry under Equilibrium Conditions at Elevated Temperatures</u>. Relative changes in stoichiometry under equilibrium conditions at elevated temperatures may also be determined from measurements of properties which are sensitive to composition. For example, the above method of combining the electrical conductivity and Hall mobility data with a simple defect model to determine composition of quenched specimens at room temperature may also be used at elevated temperatures to determine deviation from stoichiometry. Unfortunately, Hall measurements on ceramics at elevated temperatures are extremely difficult because the Hall voltage is very small. In fact, with the exception of electrical conductivity, most physical properties (lattice-parameter, magnetic, optical, etc.) are extremely difficult to measure at elevated temperatures. Thus these methods are seldom used at high temperatures to determine deviations from stoichiometry.

Changes in nonstoichiometry at elevated temperatures, however, may be studied directly through thermogravimetry by measuring the weight change of a specimen as a function of temperature and partial pressure of the more volatile component. Because of the limited accuracy of thermogravimetric techniques such studies have been limited to nonstoichiometric compounds which exhibit appreciable deviation from stoichiometry. Most of these studies have been made on metal oxides (e.g., CoO [61,62], TiO_{2-x} [63,64], Nb_2O_5 [35,65], NiO [66], $BaTiO_{3-x}$ [67]).

A schematic diagram of a typical thermogravimetric system is shown in Fig. 13. Many different types of balances may be used for

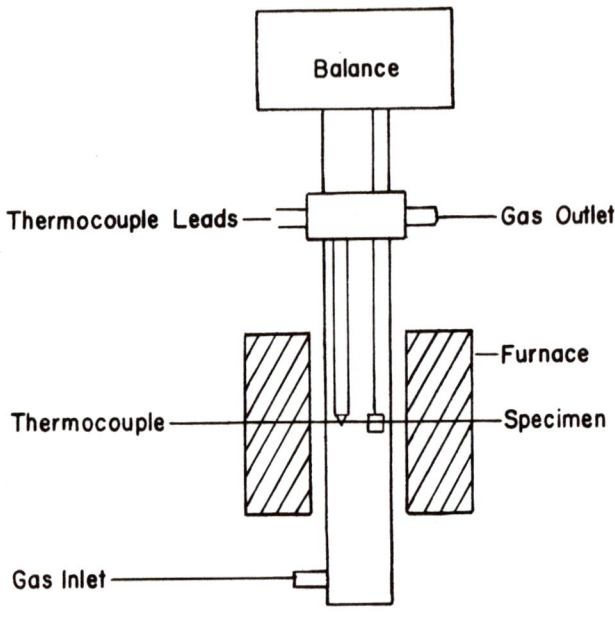

FIG. 13. Schematic diagram of a thermogravimetric system.

the weight measurements (e.g., helium balance, torsion balance), but most recent studies have been made with commercially available microbalances that have automatic recording systems [68]. Tripp et al. [49] have described systems for measuring weight changes of nonstoichiometric oxides under controlled oxygen partial pressures to 1800°C. Although the weight-change measurements appear to be simple in principle, great care must be taken to eliminate various weighing errors [69-76] caused by convection currents in the system, buoyancy forces, thermomolecular flow, vibrations, static electricity, adsorption and desorption, etc. To reduce noise from vibrations it is generally advisable to avoid wall mounting of the balance. If possible the measuring system should be located on the ground floor of the building. Commercially available platforms designed to damp out vibration may also be useful for mounting the balance system. The major problems, however, appear to arise from convection, buoyancy, and thermomolecular forces. The effect of convection appears as a noise problem (i.e., on recording balances the apparent weight of the specimen appears to oscillate rapidly about an approximate average weight) that increases with increasing temperature gradient and pressure or density of the gas. The buoyancy force also increases with increasing pressure or density of the gas. The thermomolecular force on the specimen arises from the presence of a temperature gradient. A general review of these problems has been given by Czanderna [77]. Tripp et al. [49] have attempted to minimize these effects by operating the system at a lower total pressure. They found that for their system the optimum pressure was around 0.1 atm. In subsequent measurements using a similar system but with a different balance and a substitute technique for buoyancy- and thermal-effect corrections, they claim an accuracy of weight-change measurement (for a run lasting several weeks) of ±10 μg [78].

Panlener and Blumenthal [67] have used a different approach to reduce noise from convection. They have operated their system at a total pressure of 1 atm but have reduced the density of the gas by diluting the gas mixtures (O_2-Ar) and CO-CO_2 used to control the oxygen partial pressure with He. To avoid problems with buoyancy they used gas mixtures with the same density. Because of the problems associated with convection and thermal separation of gases, the reproducibility of weight measurements with this system is much poorer than the type of system mentioned above, which uses a partial vacuum.

For the study of metal oxides the methods that are generally used to control the oxygen partial pressure are (a) blending oxygen and an inert gas, (b) blending CO and CO_2, (c) blending H_2 and water-saturated inert gas, and (d) reducing the total pressure of oxygen in the sample chamber. These methods have been described above in the section on gas-equilibration techniques.

The procedure which is generally used to measure changes in stoichiometry is described briefly below for metal-oxide systems.

According to Tripp [49] the type of sample-suspension fiber used was found to influence that amount of mechanical noise transmitted to the balance. Five-mil-diameter tungsten wire was used from the balance to the furnace tube, and depending on the temperature either a stiff quartz fiber or a sapphire rod was used in the hot zone. A reference gas mixture is generally used to define the reference state of the specimen. This state is usually selected so that the composition of the specimen is close enough to stoichiometry to be considered stoichiometric. The weight at the reference state is recorded with the gas flow-off. A gas mixture with a different value of P_{O_2} is then introduced into the system with a flow rate of approximately 50-300 cm^3/min at STP. The weight change is recorded as a function of time, when the new equilibrium state is obtained the weight of the specimen remains constant. The gas flow is then shut off and the weight change of the specimen relative to the reference state is measured. The departure from stoichiometry is then calculated from the per cent weight change of the specimen. The procedure is then repeated for other values of P_{O_2} and temperature until the departure from stoichiometry of the compound is determined as a function of P_{O_2} and temperature.

For some nonstoichiometric compounds or over certain temperature regions for a given compound, there is no gas mixture which can be assumed to correspond to the stoichiometric state. The same procedure may be used as described above for these compounds (e.g., MX_{1-x}), but the absolute value of x cannot be determined using the above method only. Only the relative change in x between the reference state and nonstoichiometric state may be determined.

Tripp et al. [78] have described a method of analyzing the weight-change data based on fitting the data to a simple defect model. The absolute value of x may be determined using this method of analysis if the appropriate defect model is used.

The above thermogravimetric technique for determining the deviation from stoichiometry is based on the assumption that only one component of the compound is volatile. For example, only oxygen is assumed to be exchanged between the compounds TiO_{2-x}, CeO_{2-x}, $BaTiO_{3-x}$ and surrounding gaseous atmosphere. This assumption should always be verified by testing the reversibility of the weight change of the specimen between two different states. For example, the thermogravimetric study of NiO reported by Tripp and Tallan [66] was limited to the temperature range 800-1100°C, because above 1100°C the permanent weight loss of the NiO specimen arising from vaporization of Ni became significant.

Some additional problems that may be encountered. (a) Balance zero may drift (i.e., the apparent weight of the specimen may drift with time); this problem can be circumvented if the measurements can be made more rapidly. (b) Static charges on the suspension wires may cause an attraction to the wall of the furnace tube, which would affect the weight measurements; a colloidal carbon solution may be used to coat the outside of the nonmetallic portions of the external system to minimize problems arising from static charges. (c) Material may condense on the sample or the hangdown tube.

Recently a tensi-volumetric method has been employed by Greskovich and Schmalzried [79] to measure oxygen nonstoichiometry in Co_2SiO_4 and in $CoAl_2O_4$-$MgAl_2O_4$ crystalline solutions between oxygen partial pressures of 2×10^{-1} and 10^{-4} in the temperature range 1100-1150°C. The schematic representation of the tensi-volumetric apparatus used by Greskovich and Schmalzried is reproduced in Fig. 14. A highly sensitive micromembrane manometer with a pressure range of 2×10^{-4} to 2×10^{-1} was used as the measuring device. The following procedure was used to measure the change in oxygen content of the ternary oxide. A sample of known weight was sealed in a quartz tube and equilibrated in dry air at known temperature and pressure. Then the oxygen partial pressure was changed to a predetermined value in the overall system. By opening and closing the appropriate valves the two identical fused silica tubes, one of which contained the sample, were interconnected via the micromembrane manometer. This procedure took about 15 sec. The starting point was found by extrapolation. Equilibration of the nonstoichiometric

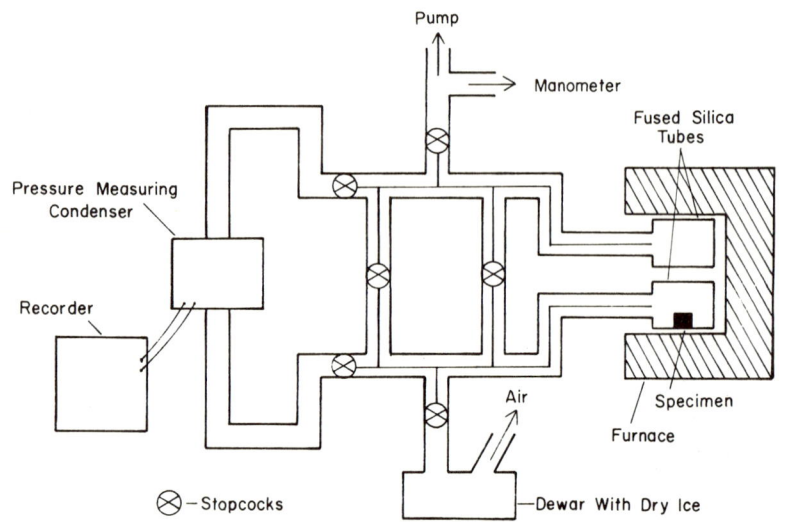

FIG. 14. Schematic diagram of the tensi-volumetric apparatus used by Greskovich and Schmalzried [79].

oxide in the lower-oxygen-pressure environment occurred by the release of oxygen from the sample, which in turn resulted in a pressure increase on the side of the system in which the sample was placed. The change in pressure ΔP was recorded as a function of time. Equilibrium was assumed when no further change in pressure occurred with time. Assuming the gas behaves according to the ideal gas law, the change in pressure caused by a change in the number of moles of gaseous oxygen (Δn_{O_2}) can be expressed by the relation

$$\Delta P = \frac{RT}{V} \pm \Delta n_{O_2} = K \Delta n_{O_2} \qquad (11)$$

In order to determine the exact value of the constant K for the system at a given furnace temperature, a $NiO_{1+\gamma}$ single crystal was used as a standard calibration specimen because it was a well known oxygen-excess semiconductor. The following equation was used for the calculation of the changes in oxygen content, $\Delta\delta$, in $Co_{2-\delta}SiO_4$ between two different values of P_{O_2} (i.e., ΔP)

$$\Delta\delta = \frac{2}{K} \cdot \frac{M_{Co_2SiO_4}}{m_{Co_2SiO_4}} \cdot \Delta P \qquad (12)$$

EXPERIMENTAL TECHNIQUES

where $M_{Co_2SiO_4}$ and $m_{Co_2SiO_4}$ are the molecular and actual weight of Co_2SiO_4, respectively.

Since this method gives only the change in the value $\Delta\delta$, two different methods were used to calculate the absolute value of δ corresponding to given temperature and P_{O_2}. For the system $Co_{2-\delta}SiO_4$ at 1145°C, δ varies between approximately 9×10^{-4} at $P_{O_2} = 2 \times 10^{-1}$ to 3×10^{-5} at $P_{O_2} \simeq 10^{-4}$. The tensi-volumetric technique employed by Greskovich and Schmalzried, although limited to a narrow range of oxygen pressure (2×10^{-1} to 2×10^{-4} atm), does have the advantage of being very sensitive to small changes in stoichiometry.

IV. ELECTRICAL CONDUCTIVITY

A. NOISE CONSIDERATIONS

Electrical-property measurements on ceramics are often beset by various complicating factors. Noise voltages may be introduced between the measurement terminals, surface conduction paths might provide a short circuit between the measurement terminals, or large resistances can exist at the electrode-specimen junction. Along with the above, instrumentation errors must also be considered.

Noise may be defined as any undesired signal that is introduced into the measurement system. The problems that noise presents are dependent on the parameter being measured and the nature of the system used for the measurement. Noise signals can either be generated internally within the specimens being investigated [80-90] or introduced from external sources [90-92]. Thermal emf's and Johnson noise are examples of internal sources of noise; electrostatic and inductive pickup, ground loops, and power-line transients are examples of external noise sources.

Thermoelectric voltages can exist when the specimen being studied is in a temperature gradient and thus may be minimized by insuring that the specimen is in a constant-temperature environment.

When this is impractical the thermoelectric contribution to the measured signal may be eliminated by performing bidirectional measurements and averaging the results obtained from each polarity of measurement. It is noted that ac techniques rapidly accomplish bidirectional measurements.

Johnson or Nyquist noise is due to thermal agitation and can be quite troublesome when one is taking measurements at high temperatures on materials having high resistivities. In a homogeneous material the rms Johnson-noise voltage E_r can be expressed as

$$E_r = \sqrt{4kT \Delta f R} \tag{13}$$

where k is Boltzmann's constant, T is the absolute temperature, R is the resistance between the voltage-measuring terminals and Δf is the bandwidth of the measurement. This noise exhibits itself as random voltages of various magnitudes and frequencies and may be minimized with the aid of bandwidth filters. If measurements are being performed at very high temperatures, then one might have to consider the shorting effect of thermionically emitted electrons from the surface of the specimen and specimen holder. For intermediate frequencies generation-recombination noise due to fluctuations in the density of free carriers as a result of background radiation can exist in photosensitive materials. At low frequencies, "1/f" noise associated with surfaces, barriers, and poor contacts might also exist.

The most familiar source of noise external to the specimen is electrostatic pickup. This is quite troublesome in high-impedance circuitry. When the distributed capacitance C between the measurement leads has a charge Q impressed upon it, a voltage

$$V = Q/C \tag{14}$$

is developed across these leads. The most effective means of minimizing this source of noise is to shield the measurement circuitry from stray sources of charge. A shield is composed of a highly conductive material, such as copper screening or foil, placed around

EXPERIMENTAL TECHNIQUES

the measurement system and its leads. This shield is grounded so that any stray charge is shorted to ground. It is important that all grounding be at the same potential in order to eliminate voltage drops introduced when current flows from a ground at one potential to a ground at a different potential (ground loops). Grounded shields are considered good practice and should be included in all systems to be considered.

Inductive pickup from surrounding power sources can also lead to noise introduction. The source signal might result from power distribution systems, from radio-frequency transmission equipment, or from arcing. The use of short leads and twisted pairs of signal leads will aid in minimizing this source of noise.

Transient signals introduced from ac power lines are often found to lead to noise. These may be suppressed by the use of power-line filters and isolation transformers. If the source of transient noise can be located it may be eliminated; otherwise a different power line should be used. The use of battery-operated instrumentation would also aid in removing this source of noise.

B. EXPERIMENTAL METHODS

The electrical conductivity of a ceramic material is an important property that can be used to characterize the internal structure of a material or aid in determining its transport properties. In order to use this parameter as a means of analysis, measurements must be performed on as wide a range of materials as need to be characterized. The conductivity of ceramic materials can vary over extreme ranges with adjustments of the external variables. Numerous methods are available to measure the conductivity of a material, each having their own merits. A summary of the experimental methods for the measurement of conductivity is given in Table 4. These methods are now discussed individually.

TABLE 4

Experimental Methods for the Measurement of Conductivity

Method	Remarks
Two-probe dc method	This method is simple to use, but ohmic contacts must exist if measurements are to be representative of the specimen's bulk behavior. This method is often the only practical one when investigating very-high-resistivity materials.
Two-probe ac method	Varying the frequency of measurement can aid in eliminating junction effects and in distinguishing ionic from electronic current flow.
Three-probe ac and dc methods	Three-probe methods can be used to distinguish surface and bulk conduction processes in materials having low to moderate conductivities. Ohmic electrodes are again required if values are to be representative of the bulk material.
Four-probe methods	The four-probe methods are best suited to materials having moderate to high conductivities. With this method the effects of nonohmic contacts or barrier layers can be overcome. Certain four-probe methods are useful when working with irregularly shaped specimens.
Multi-probe methods with 5 or more probes	These methods are useful in situations where nonlinear fields exist along the specimens being investigated.

The basic configuration for the two-probe conductivity method is shown in Fig. 15. A constant voltage is impressed across the specimen, or a constant current is passed through the specimen. The voltage V across and the current I through the specimen are measured, and the conductivity σ is calculated using

$$\sigma = \frac{dI}{AV} \tag{15}$$

EXPERIMENTAL TECHNIQUES

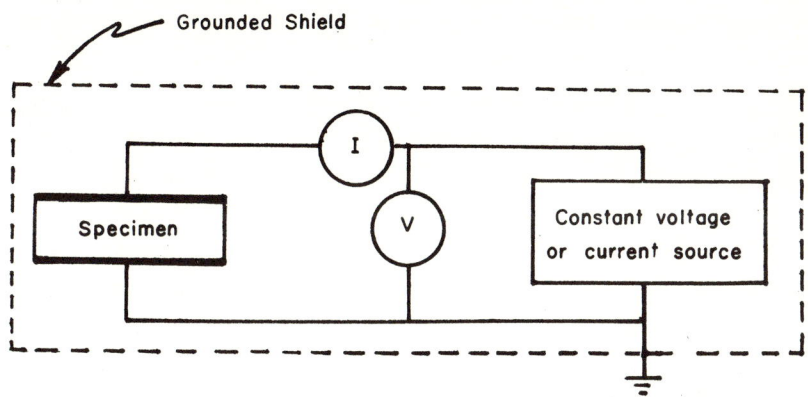

FIG. 15. Circuit for the two-probe measurement of the dc electrical conductivity.

where d is the specimen thickness or distance between the two electrodes and A is the cross-sectional area perpendicular to the current direction in the sample. Equation (15) is a statement of Ohm's Law which is applicable only where a homogeneous material exists between the electrodes. The reliability of the measured conductivity will depend on numerous factors, including the ranges of conductivity and the instrumentation being used. The voltage drop across the current-measuring device must be negligible compared to that of the specimen. If an ammeter is to be used, its internal resistance must be small. If the current is determined by measuring the voltage drop across a standard resistor, the resistance of this standard must be small compared with that of the specimen. Where the resistance of the current-measuring unit is large, a correction is required for the voltage drop across it [92]. Conversely, the impedance of the voltage-measuring device must be very large with respect to the resistance of the specimen between its electrodes. If this condition is not met, then a correction might have to be made for the current flowing through this instrument [92].

The experimental problems associated with very-low-conductivity materials [$\sigma \lesssim 10^{-8}$ ohm^{-1} cm^{-1}] will be considered first. With this type of material, one must insure that the current path is through the bulk material between the electrodes. Since current flows in the path of least resistance, shorting paths might exist along the surface of the sample, through the specimen holder, or through the surrounding atmosphere. One must make sure that the percentage of current flowing in the shorting paths is less than the experimental error desired in the data; i.e., if the conductivity is to have an error of less than ±1%, less than 1% of the total current should be allowed to flow through extraneous current paths. The extraneous current path that is most difficult to take into account is that associated with surface conduction. One can minimize this source of error and still use the two-probe method by making the bulk path length for current flow much shorter than the surface path length, as shown in Fig. 16 [93]. It is noted that the effect of the other two sources of surface conduction can be determined by making a measurement of the current flowing through the system before (I') and after (I'') the specimen has been placed in the holder. If $I' \ll I''$ these two sources of leakage are negligible. As is discussed, other multiprobe techniques are better adapted for protection against leakage paths. When working with low-conductivity materials the voltage-measuring circuitry will usually be working into high impedances, and, therefore, shielding will be needed to reduce electrostatic pickup.

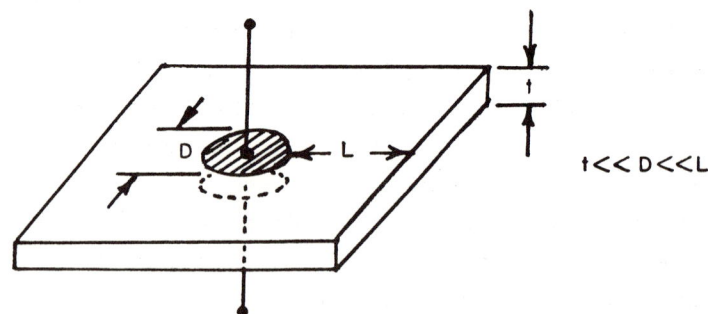

FIG. 16. Electrode arrangement to minimize contributions from surface conduction.

EXPERIMENTAL TECHNIQUES

Surface conduction is usually not a problem with high-conductivity specimens. However, voltage drops across the electrode-specimen junction can be troublesome. High-resistance contacts can be the result of barrier layers or nonohmic behavior of the electrodes. It is noted that this might also occur for low-conductivity materials. If two-probe dc conductivity measurements are to be reliable, one must be sure that the contacts are ohmic and that no barrier layers exist. If this cannot be established, then multiprobe or ac measurements should be conducted.

In dealing with ceramics which have both electronic and ionic conductivity contributions added precautions must be taken. In order to measure the true bulk conductivity in a specimen of this type using two-probe dc techniques the electrodes have to be nonblocking to both the electronic and ionic species. If this precaution is not taken, a space charge of either ions or electrons will result in the vicinity of the electrodes. This leads to a nonuniform field across the specimen, and Ohm's Law as stated in Eq. (15) does not apply.

In many situations a two-probe ac technique can prove of value when barrier layers exist at the contacts or when appreciable ionic conductivity is present. A schematic representation of this type of system is shown in Fig. 17. Many commercial impedance bridges are available, and generally suppliers are quite helpful in selecting the unit required. Ratio-transformer-type bridges seem to offer the widest versatility of usage and are able to perform measurements over wide impedance ranges, although their frequency ranges are sometimes limited. In choosing a variable-frequency source a low-output impedance is desired, since these instruments have better voltage regulation under such conditions; i.e., the output voltage remains constant when working into lower-impedance circuitry. The selection of a tuned null detector is quite important. This detector should have a high sensitivity over the desired frequency range and should be able to detect voltages in the microvolt to tenth-of-microvolt regions. These detectors should have a high value of Q,

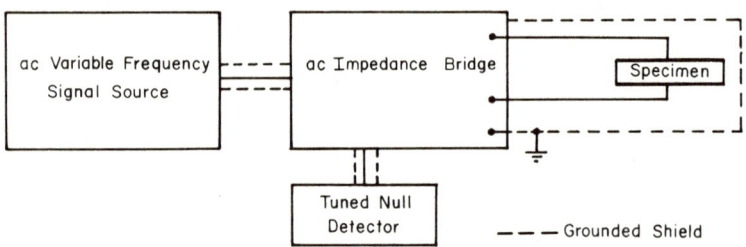

FIG. 17. Circuit used to measure the two-probe ac electrical conductivity.

i.e., they should be capable of performing measurements over a narrow bandwidth centered around the measuring frequency; this property aids in eliminating the effects of spurious signals.

When the investigator is only interested in measuring the conductivity of his specimen the two-probe ac method may fulfill his requirement with moderate effort. As the frequency of measurement is increased, the effects of barrier layers in the vicinity of the electrodes will be eliminated. This can be envisioned with the aid of Fig. 18.

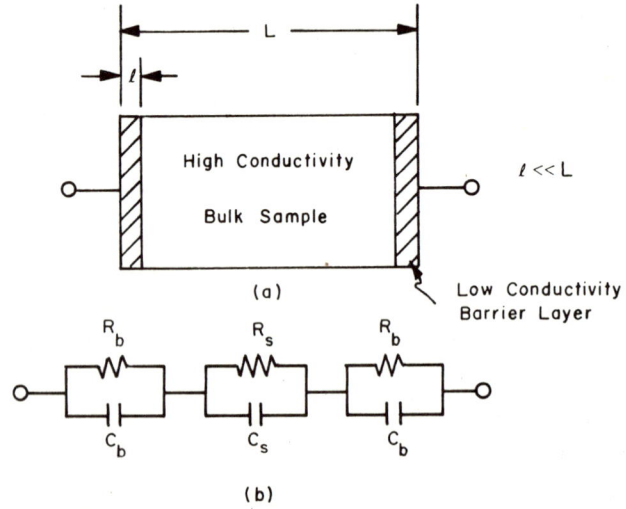

FIG. 18. Specimen (a) having thin, high-resistivity barrier layers, and (b) its circuit representation.

EXPERIMENTAL TECHNIQUES   101

   Figure 18 shows a specimen which has barrier layers in the vicinity of its contacts. These are usually thin compared with the length of the bulk specimen. If the resistance R_b of these barrier layers is very large compared to the bulk resistance R_s it is observed that at low frequencies the major voltage drop between the two electrodes occurs across these barrier layers. The configurational capacitance C of a specimen is given by

$$C = \frac{\epsilon' A}{d} \qquad (16)$$

where ϵ' is the permittivity of the material composing the specimen, which has area A and thickness d. The capacitance of the barrier layer would be

$$C_b = \frac{\epsilon' A}{\ell} \qquad (17)$$

and of the bulk material

$$C_s \simeq \frac{\epsilon' A}{L} \qquad (18)$$

Since ℓ is much less than L, the capacitance C_b of the barrier layer will be much greater than that associated with the bulk specimen C_s. Therefore, as the measurement frequency is increased, the barrier layer will tend to be short circuited by virtue of its reduced capacitive reactance. There will be a frequency at which any contribution from barrier layers in the vicinity of the contacts would be eliminated.

   When one is working with materials which have both electronic and ionic contributions to the total conductivity, space-charge polarization at the electrodes can become appreciable under dc or low-frequency ac conditions. At higher measurement frequencies polarization at the specimen electrodes does not have sufficient time to build up and the effect of electrode polarization may be eliminated.

   When one is performing bridge-type measurements the specimen capacitance and dielectric constant can be measured simultaneously with its conductance and conductivity. This is easily done when one is working with low-conductivity materials, but is quite difficult to do with high-conductivity materials. Impedance bridges, if they are to be used for capacitance measurement, are able to measure capacitance on specimens having dissipation factors up to about 10^1. The dissipation factor is the ratio of the specimen conductance G to capacitive admittance ωC. If capacitance measurements are to be performed on highly conductive materials it is desirable for one to be able to perform measurements on materials having dissipation factors as large as 10^5. If this is to be achieved, various stray inductances and capacitances in the measurement system must be taken into account [94].

A system that is able to accomplish this is shown in Fig. 19. Figure 19(a) shows a schematic of the measurement system; Fig. 19(b) depicts the specimen holder; and Fig. 19(c) shows a guarded fixture which is discussed later. Three separate measurements are performed. For the first measurement the capacitance C_1 of the circuit with the lead removed from the upper electrode at point D [see Fig. 19(b)] is measured with the guarded leads of the bridge (A and B) connected to terminals A' and B' of the specimen holder, and is

$$C_1 = C_{sh} \tag{19}$$

where C_{sh} is the capacitance of the leads up to the specimen electrode. The specimen (with its electrodes) is then connected into the circuit by connecting the upper lead, and the bridge is balanced by adjusting the capacitance of the bridge and the external resistance provided by the decade boxes. The value of capacitance C_2 read from the bridge is

$$C_2 = C_{sh} + C_{sp} - MXC_{db} + A_{db} - B_1 \tag{20}$$

where C_{sp} is the specimen capacitance; MXC_{db} is the capacitance of the decade box (C_{db}) reflected through the bridge circuit and multiplied by factors M and X associated with the bridge; A_{db} is a factor associated with the inductance of the decade box, which is

EXPERIMENTAL TECHNIQUES

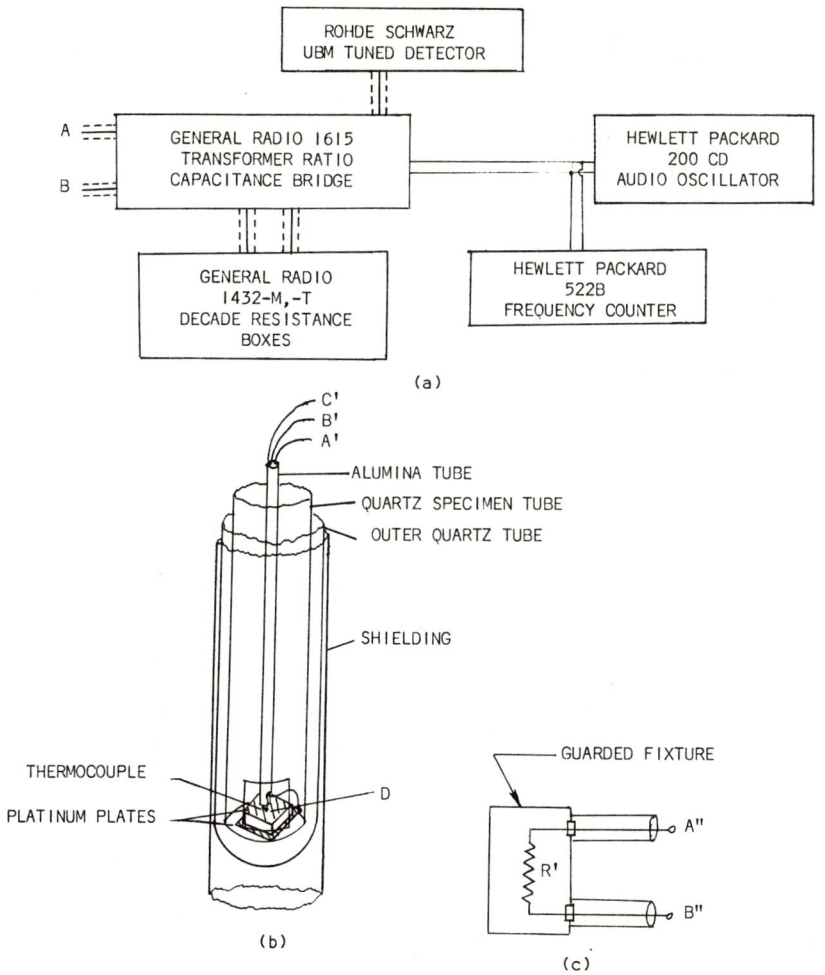

FIG. 19. System capable of measuring capacitance on materials having a dissipation factor as high as 10^5. (a) Schematic of the experimental apparatus, (b) specimen holder (A′-B′—specimen leads; B′-C′—thermocouple leads), (c) guarded fixture.

reflected through the bridge; and B_1 is a term associated with the lead inductance of the specimen holder. The resistance of the specimen is given by

$$R_{sp} = \frac{R_{db}}{MX} \qquad (21)$$

where R_{db} is the resistance of the decade box. After these readings, the guarded terminals of the bridge A-B are connected to terminals A″-B″ of the guarded fixture [shown in Fig. 19(c)]. The resistance R′ used in this guarded fixture is chosen to be the same as that of the specimen as calculated from Eq. (21). The bridge is then rebalanced by varying the capacitance of the bridge circuit only. The capacitance reading C_3 is given by

$$C_3 = C_r - MXC_{db} + A_{db} \tag{22}$$

where C_r is the capacitance of the resistor R′ and the fixture holding it. If one chooses a resistor which has a small capacitance and if the specimen leads are designed to have a small inductance, the terms C_r and B_1 are small. The specimen capacitance may thus be given by

$$C_{sp} = C_2 - C_1 - C_3 \tag{23}$$

The two-probe dc method of measuring specimen conductivity was found to be simple in implementation. However, it was inadequate in dealing with mixed conduction, barrier layers, and surface conduction. The first two of these inadequacies could be somewhat alleviated by using the two-probe ac method. In order to deal with the problems associated with surface conduction, three-probe methods must be considered [95-100]. The purpose of the third probe in both the dc and ac methods is identical and may be discussed simultaneously.

The basic electrode configuration for three-probe measurements on plane parallel specimens is shown in Fig. 20 [95]. The third electrode, electrode a, is referred to as either the guard electrode or the guard ring. This guard electrode is often connected to the grounded shield, and therefore is at ground potential; or it is at the same potential as is electrode b. The remaining two electrodes, b and c, are used for the measurement of the bulk specimen conductivity. Any surface current that exists in the specimen would be prevented from having surface conduction path between b and c.

EXPERIMENTAL TECHNIQUES   105

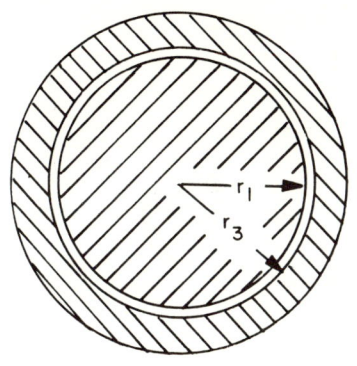

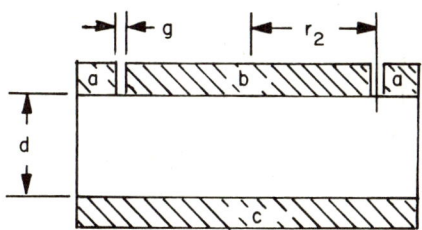

FIG. 20. Electrode configurations for bulk and surface conductivity measurements on plane parallel specimens. $r_2 = \frac{1}{2}(r_1 + r_3)$; $g \leq 2d < R_1$.

Surface conduction is most often a problem when studying low-conductivity materials. This is because the surface is the path of least resistance. Volt-ampere or bridge techniques may be used to measure the resistance between electrodes b and c. Typical circuits that can be used for this purpose are shown in Fig. 21 [90, 95-97]. If the resistance is measured between electrodes b and c the bulk conductivity may be calculated from the equation

$$\sigma_b = \frac{d}{RA} \tag{24}$$

where $A = \pi(2r_1 + g)/4$ is the effective area of electrode b. It is observed that one might also obtain the capacitance between electrodes b and c when the bridge method is used.

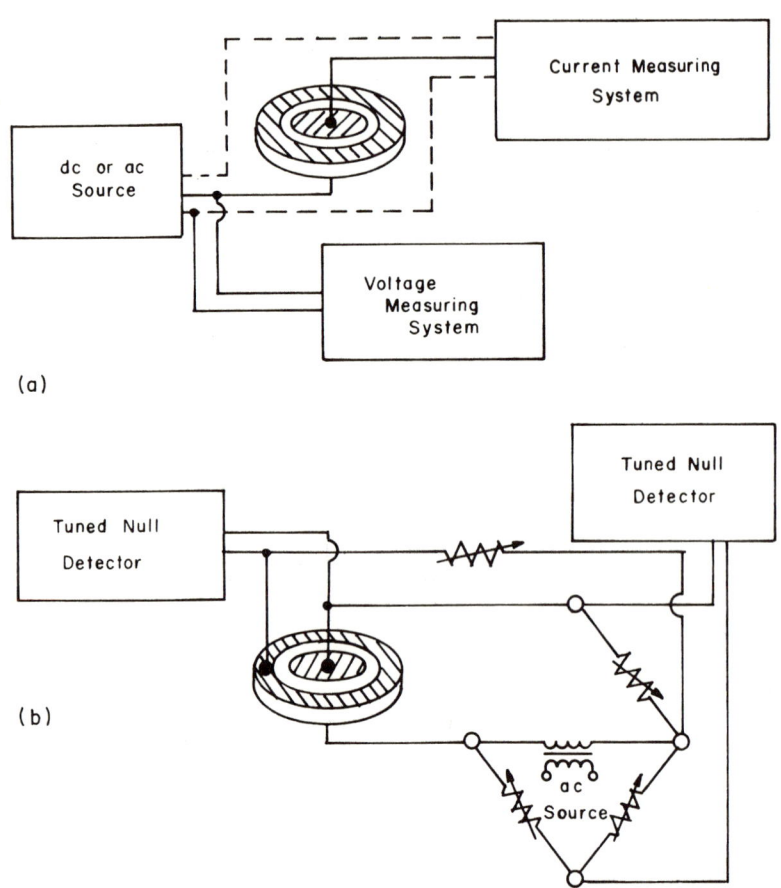

FIG. 21. Circuits for three-probe measurements of bulk specimen conductivity. (a) Volt-ampere method; (b) bridge method.

The three-probe configuration can also be used to measure the surface conductivity. A circuit for this purpose, using a volt-ampere method, is shown in Fig. 22 [90, 95-97]. If the resistance R_s is measured between electrodes a and b, then the surface conductivity is given by

EXPERIMENTAL TECHNIQUES														107

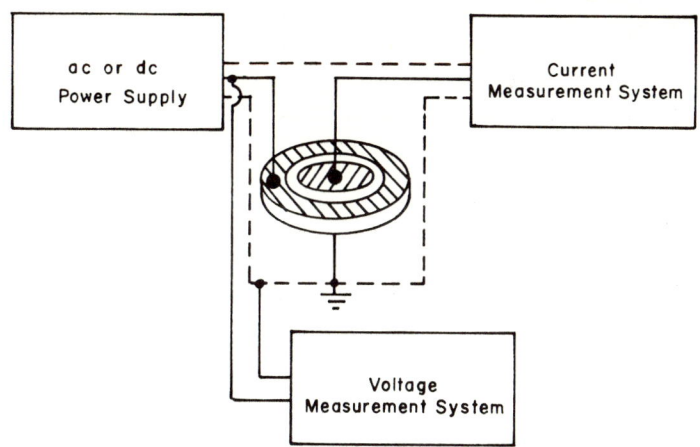

FIG. 22. Circuit for three-probe volt-ampere method of measuring surface conductivity.

$$\sigma_s \mp \frac{g}{R_s\, 2\pi(r_1 + g)} \tag{25}$$

When specimen configurations other than that shown in Fig. 20 are used other equations must be used to calculate the bulk and surface conductivities [90, 95-97].

Four-probe conductivity techniques have been developed which alleviate a number of problems that might be encountered in practice [101-104]. These generally are found to function best on materials having low to moderate resistivities and give rise to many difficulties when used with high-resistivity materials. A basic system that might be used is shown in Fig. 23. Although this is not shown, the measurement system and leads should, again, be enclosed by a grounded shield. If the current I through the specimen and the voltage v across the two inner probes are measured the specimen conductivity is given by

$$\sigma = \frac{I\ell}{VA} \tag{26}$$

where A is the area of the specimen and ℓ is the separation of the two inner probes. No portion of the voltage drop V occurs across

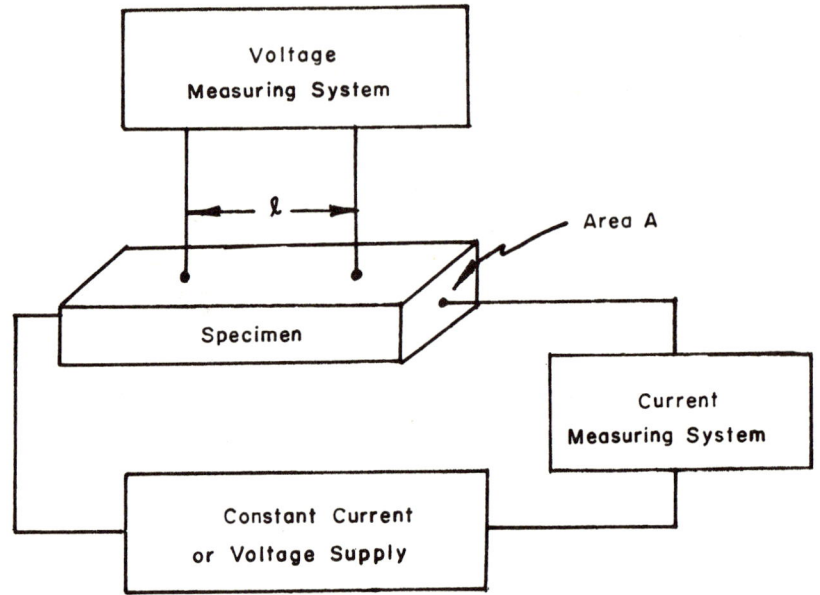

FIG. 23. System used for the four-probe conductivity measurement.

the outer current contacts and, therefore, any barrier layers in the vicinity of these contacts do not affect the measured value of conductivity. The input impedance of the system used to measure the voltage V should be as high as possible. If no current is allowed to flow in the voltage-measurement leads, there will be no voltage drop across the resistance associated with the voltage contacts and this resistance will not influence the conductivity measurement. Note also that the condition where the input resistance of the voltage-measurement system is very much larger than that between the measurement leads becomes more difficult to achieve when working with high-resistivity specimens.

For very-low-resistivity specimens the voltage sensitivity of the measurement system would have to be sufficient to measure the small voltage drop occurring across the voltage probes. This voltage drop might be increased if the current is increased, but an upper limit would be reached due to Joule heating of the specimen.

EXPERIMENTAL TECHNIQUES   109

The above discussion is applicable to both dc and ac four-probe conductivity measurements. In ac conductivity measurements there is an added problem due to the capacitance between the various leads in the measurement system. If the conductivity is measured as a function of frequency and if it is found that after a certain frequency the conductivity increases with increasing frequency in an unlimited fashion, as is shown in Fig. 24, then the experimenter must be concerned with the possibility that the voltage-measurement leads are being shorted due to inter-lead capacitance. If this is the case, either the frequency of measurement will be limited by this condition or guarded voltage-measurement leads will be required.

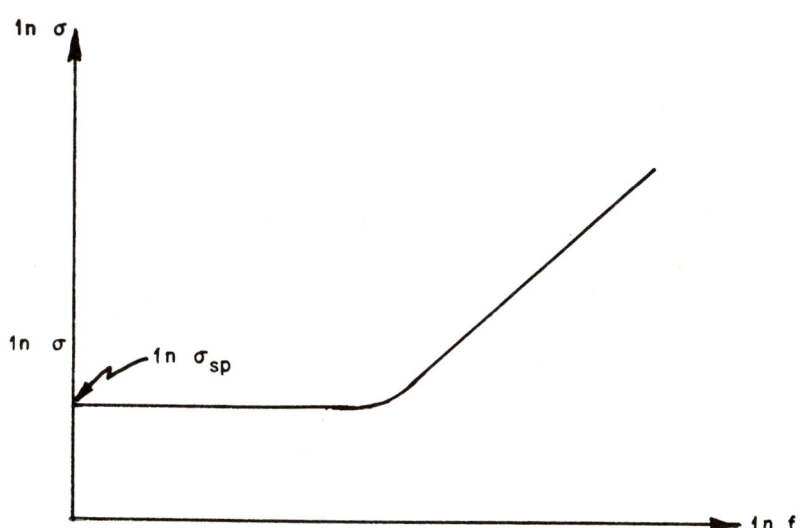

FIG. 24. Plot of conductivity vs frequency when capacitive coupling occurs between the voltage probes.

A four-probe technique as developed by van der Pauw is found to be useful in working with flat samples of arbitrary shapes, as shown in Fig. 25 [104]. The sample is assumed to be of uniform thickness d with no holes or asperities within its circumference. Small contacts are placed on the circumference of the sample and are designated A, B, C, and D. Two values of resistances are

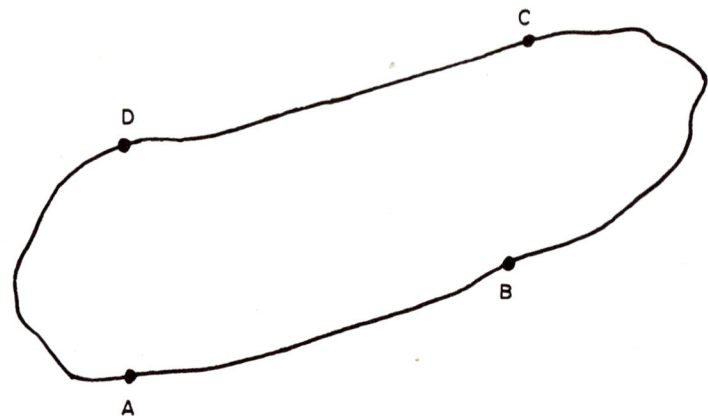

FIG. 25. Electrode arrangement used in the van der Pauw method.

measured. The first is $R_{AB,CD}$ which is defined as the ratio of the potential difference developed between the terminals CD to the current passing through the terminals AB. The second is $R_{BC,DA}$, which is the ratio of the potential difference developed between the terminals DA to the current passing through terminals BC. It has been shown [104] that

$$\exp \frac{-\pi d R_{AB,CD}}{\rho} + \exp \frac{-\pi d R_{BC,DA}}{\rho} = 1 \qquad (27)$$

where ρ is the resistivity of the specimen being considered. It has been further shown that

$$\rho = \frac{\pi d}{\ln 2} \frac{R_{AB,CD} + R_{BC,DA}}{2} \cdot f\left(\frac{R_{AB,CD}}{R_{BC,DA}}\right) \qquad (28)$$

where $f(R_{AB,CD}/R_{BC,DA})$ is a function which depends on the ratio $R_{AB,CD}/R_{BC,DA}$, as is shown in Fig. 26.

Van der Pauw has further shown that the Hall mobility can be determined by measuring the change in the resistance $R_{BD,AC}$ with the application of a magnetic field perpendicular to the parallel surfaces of the sample. The Hall mobility is given by

EXPERIMENTAL TECHNIQUES   111

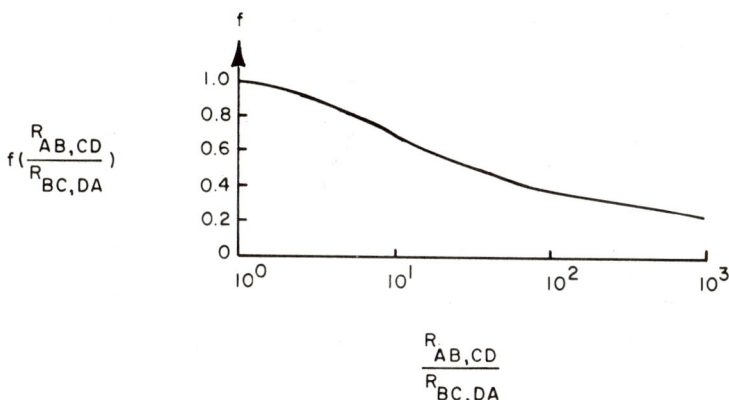

FIG. 26. Plot of function used by van der Pauw for conductivity calculations.

$$\mu_H = \frac{d}{B} \frac{\Delta R_{BD,AC}}{\rho} \tag{29}$$

where B is the magnetic-field strength and $\Delta R_{BD,AC}$ is the change of the resistance $R_{BD,AC}$ upon the application of the magnetic field.

Other four-probe techniques have been reported in the literature that are useful when working with thin-layer specimens [102,103]. These depend on very accurate probe alignment and spacing, but under some circumstances they are useful when working with ceramics. These would be of particular interest when working with samples having conducting boundaries along the circumference or on one of its plane parallel surfaces.

A six-probe conductivity-measurement technique has been employed by Blumenthal et al. [105] to study conductivity variations along bar samples under the influence of an electric field. A schematic of the system used is shown in Fig. 27. A constant current is impressed through the outer contacts and the voltages across the various probe pairs are recorded as a function of time using a high-input impedance multichannel oscillograph. Knowing the current density $J = 1/A$ flowing through the specimen and the voltage V across the various probe pairs that are separated by spacing ℓ, one may find the

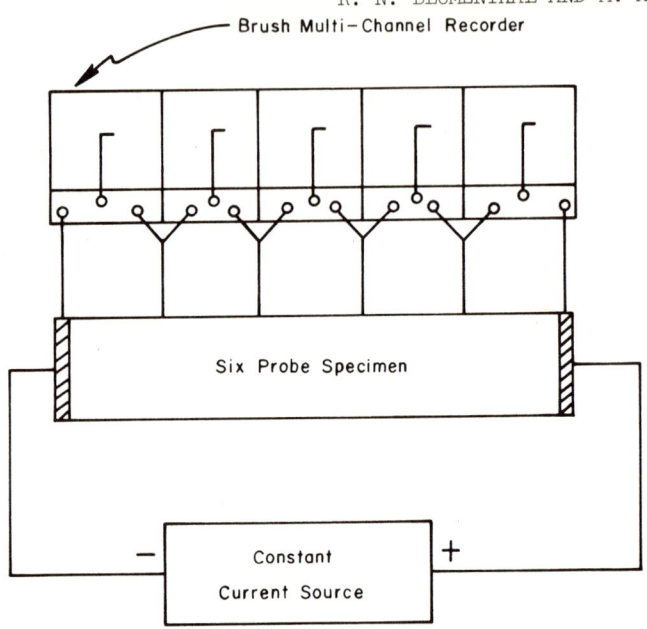

FIG. 27. Six-probe conductivity system used by Blumenthal et al. [105].

conductivity variations along the specimen as a function of time. This method proves quite useful in dealing with situations where nonlinear fields develop along the length of the specimens under the influence of an applied field.

V. TRANSFERENCE MEASUREMENTS ON MIXED CONDUCTORS

In a compound exhibiting mixed conduction both the electronic and ionic conductivities are functions of the concentration of electronic and ionic defects. The type and concentration of defects are generally very dependent on the deviations from stoichiometry and the amount of impurities or dopant. In order to obtain meaningful values of the transference numbers it is necessary to make these measurements under conditions where the composition of the compound is well defined. For a binary compound this can be accomplished by equilibrating the sample with a gaseous atmosphere or solid in which

EXPERIMENTAL TECHNIQUES

the thermodynamic potential of one of the components of the sample is fixed. The solid is often chosen to act as an electrical contact.

Wagner [106] and Hebb [107] have made the observation that the fraction of current carried by ions depends on the type of electrode used. According to Kroger, four types of electrodes may be distinguished [108]:

- Type a: reversible electrodes
- Type b: electrodes passing only electronic currents (semiblocking electrodes)
- Type c: electrodes passing only ionic currents (semiblocking electrodes)
- Type d: electrodes passing neither electronic nor ionic currents (blocking electrodes)

Wagner has derived equations for cells involving combinations of the above types of electrodes [109]. In this analysis he considers only transport of cations, anions, and electrons. Kroger has also derived equations for the same type of cells [108]. This analysis, however, is based on the transport of defects. In his review of Wagner's equation, Kroger [108] compares both methods of describing transport.

Since electrodes of Types a, b, or c are most frequently used in cells for determining transference numbers the discussion of experimental techniques will be confined to cells using these types of electrodes.

A. REVERSIBLE ELECTRODES

Wagner [109] has described two different types of galvanic cells with reversible electrodes that may be used to measure transference numbers. In one cell a mixed conductor is placed between identical reversible electrodes and in the other a mixed conductor is located between unlike reversible electrodes.

For a reversible electrode the electrode and the mixed conductor must have an ion in common and any reactions involving electrons and ions taking place at the interface must be free to occur (i.e., when a current is passed through a cell with reversible electrodes, there should not be any potential drop at the interface between the electrode and mixed conductor or a polarization voltage occurring in the mixed conductor). Reversible electrodes usually have two functions: (a) they define the chemical potential of one of the constituents of the mixed conductor at its interface with the electrode; and (b) they act as a current lead and or a potential probe. The above two types of cells with reversible electrodes will be described below.

1. Galvanic Cells with Identical Reversible Electrodes

Typical examples of a mixed conductor (e.g., M_aX_b) between identical reversible electrodes are represented schematically in Fig. 28. In Fig. 28(a) the compound M_aX_b is assumed to be in thermodynamic equilibrium with the metal electrode M (i.e., the value of the ratio a/b is fixed by the chemical potential of M). In Fig. 28(b) the compound M_aX_b is assumed to be in thermodynamic equilibrium with the gas (X_2)/inert-metal electrode. The chemical potential of X is controlled by the partial pressure of X_2 in the surrounding atmosphere. For either of the above two types of cells the ratio of metal to nonmetal should be uniform throughout the compound. Thus with no current flow the emf of these cells should be zero. If an external field is applied to these cells, all of the potential drop should occur across the mixed conductor. Since there is no gradient in composition or chemical potential, the gradient in the electrostatic potential should be uniform and Ohm's Law should be obeyed (i.e., the current should be proportional to the applied potential). Thus the total conductivity σ_T, which is the sum of the electronic and ionic conductivities, can be determined from the quotient of the current density divided by the electronic field. The ratio of the ionic conductivity σ_i to the total conductivity is the ionic transference number t_i:

EXPERIMENTAL TECHNIQUES

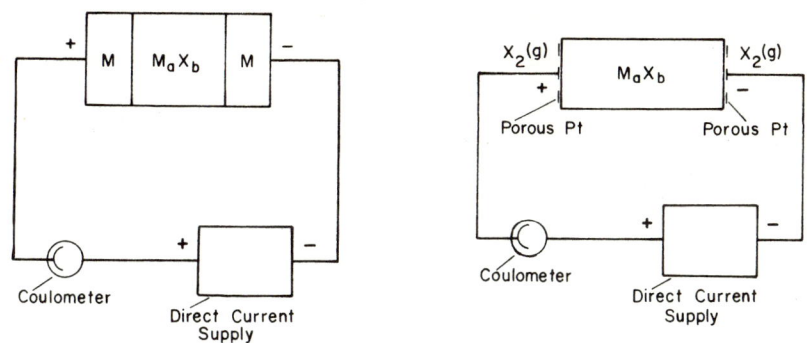

FIG. 28. Mixed ionic and electronic conductor M_aX_b between identified reversible electrodes.

$$t_i = \frac{\sigma_i}{\sigma_i + \sigma_e} \tag{30}$$

The ionic transference number is also equal to the ratio of the number of equivalents of metal or nonmetal evolved or consumed at the electrodes to the total number of faradays passed through the cell; thus

$$t_i = \frac{J_i}{J_i + J_e} \tag{31}$$

If a current is passed through the cell shown in Fig. 28(a) the cations will migrate to the cathode (i.e., the negative metallic electrode) where they will be deposited in the metallic form, and at the anode cations will be formed from the metallic electrode. The anions will migrate from the cathode to the anode where they will react with the metal electrode to form M_aX_b. At the cathode, however, the reverse reaction will occur (i.e., M_aX_b will decompose to form M and anions). Thus, the net effect of an ionic current will be to increase the weight of the cathode and to decrease the weight of the anode. One of the advantages of this method is that the number of equivalents, and hence the amount of current carried by the ions, can be easily determined from the weight change of the electrodes.

One of the disadvantages with this type of cell is that the ionic transference number of the compound can be determined for only one composition at any temperature (i.e., the composition corresponding to the equilibrium between the metal electrode and the compound).

This limitation can be avoided by the use of the cell shown in Fig. 28(b). With this type of cell the ionic current may be determined from the number of equivalents of $X_{2(g)}$ taken up or evolved at the corresponding inert electrodes. In principle, with this type of cell t_i may be determined for the compound as a function of composition by varying the partial pressure of X_2. With this technique the ionic transference number 3×10^{-6} was determined for CuI at 200°C and at iodine partial pressure of 0.06 atm [110].

Aside from the problem of using reversible electrodes, the major experimental problems with the cell of the type shown in Fig. 28(b) are (a) to obtain a gas-tight seal to separate the atmosphere at the two electrodes, and (b) to find an experimental technique that allows one to determine the amount of gas taken up or evolved at the inert-metal electrodes.

It should be noted that with either of the above cells no information is obtained about what fraction of the ionic current is carried by cations or anions. To obtain this type of data the chemical method initiated by Tubandt [111] may be used. Figure 29 is a schematic diagram of the apparatus used by Tubandt to measure the transference number of Ag in α-AgI. He placed the weighted cylindrical pellets of α-AgI (I, II, and III in Fig. 29) in series between a silver anode and a platinum cathode of known weight. A small current was passed through the cell for a period of time. The total charge was determined from the equivalent weight of silver deposited in the coulometer. All the cylinders and electrodes were separated from one another, with the exception of cylinder I and the Pt electrode, which had stuck together. These cylinders and electrodes were then reweighed. The weight of silver deposited in the coulometer

EXPERIMENTAL TECHNIQUES

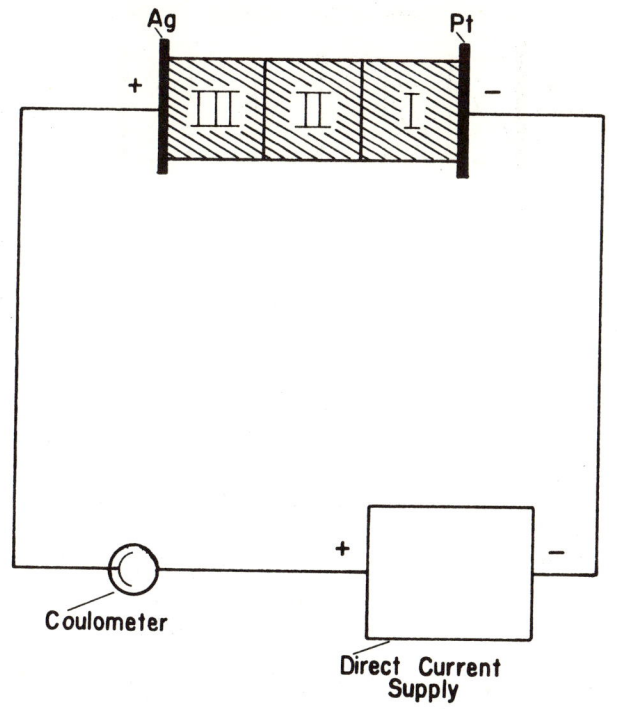

Fig. 29. Schematic diagram of the apparatus used by Tubandt to measure the ionic transference number of Ag in α-AgI.

was equal to the combined increase in weight of cylinder I and the Pt electrode and the decrease in weight of the Ag anode. Since electronic conduction would not have produced any change in weight of the electrodes and anion conduction would have resulted in an increase in weight of cylinder III and a decrease in weight of cylinder I, the transference numbers for cations may be interpreted as unity within experimental error. A problem that frequently occurs with the simple arrangement shown in Fig. 29 is the deposition of metal at the cathode in the shape of needles which grow into the cylinder and short circuit the electrodes. To circumvent this problem Tubandt used a "protective" electrolyte in series with the electrolyte being investigated. For example, he used pellets of AgI in series with AgBr, as shown in Fig. 30, to prevent the growth of

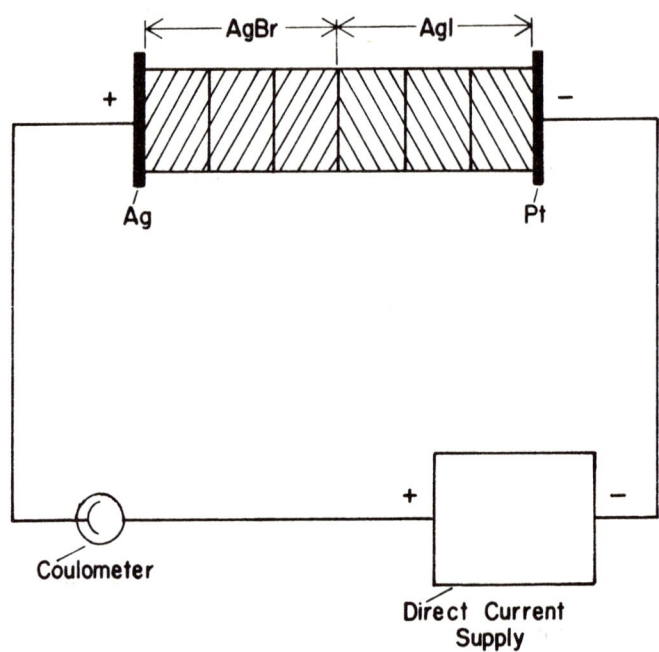

FIG. 30. Schematic diagram of the apparatus used by Tubandt to measure the ionic transference number of AgBr.

Ag needles into the AgBr. With this arrangement the Ag ions pass from the AgBr to the AgI and are deposited at the AgI-cathode junction. The bromine ions which remain at the anode react with Ag to form AgBr.

In both of the above examples the transference number for the cation was found to be essentially unity. For compounds exhibiting both cation and anion conduction it is often necessary to use a more complicated arrangement. For a mixed conductor like NaCl Tubandt [111] placed cylinders of $BaCl_2$ ($t_- = 1$) on both sides of the NaCl cylinders. Only Cl^- ions migrated through the $BaCl_2$ cylinder, whereas both Na^+ and Cl^- migrated through the NaCl cylinders. If 1 F of electricity is passed through the cell, 1 g equiv. of Cl^- ions passes through the $BaCl_2$ cylinder and arrives at the interface between the $BaCl_2$ and NaCl cylinders on the cathode side of the cell.

EXPERIMENTAL TECHNIQUES

In the NaCl cylinder adjacent to this interface, t_- of an equivalent of Cl^- ions migrate away from this interface and t_+ of an equivalent of Na^+ ions arrive at this interface. The net result is the formation of t_+ of an equivalent of NaCl at this interface. Conversely the interface between the NaCl and $BaCl_2$ cylinders on the anode side of the cell loses t_+ of an equivalent of NaCl. Tubandt [111] observed that the neighboring cylinders of NaCl and $BaCl_2$ tended to stick together after current was passed through the cell. Rather than attempt to separate the cylinders, he weighed them together. He found that the pair of NaCl and $BaCl_2$ cylinders nearest the cathode gained weight, and the pair of NaCl and $BaCl_2$ cylinders nearest the anode lost an equivalent amount of weight. The transference number t_+ for Na^+ ions was determined by dividing the above weight change at the cathode by the equivalent weight of NaCl passed through the cell, which was calculated from the total charge passed through the cell using a silver coulometer.

Transport measurement by Tubandt's method has the advantage that relatively simple measuring systems may be used. However, the transport number calculated from measurements obtained from these types of cells may be significantly in error for a number of reasons. For example, at high temperatures problems may arise from loss of weight by evaporation or from diffusion of the "protective" electrolyte into the material being studied. The use of a "protective" electrolyte may also cause blocking of current carriers (i.e., if an electrolyte with $t_i \simeq 1$ is placed in series with a conductor which exhibits both electronic and ionic conduction, the ionic electrolyte may block the electronic charge carriers and thus alter the ionic transport number of the material). For transport measurements on Ag_2S, Tubandt [111] used an experimental arrangement similar to that shown in Fig. 30, with α-AgI as a "protective" electrolyte. From the results obtained with this cell he concluded that the transference number for Ag^+ ions in Ag_2S, $t_+ \simeq 1$. Wagner [106], however, measured the transport number of Ag_2S and concluded that it was primarily an electronic conductor.

It should also be noted when "protective" electrolytes are used, as in the Tubandt-type cells, the material being studied is not in thermodynamic equilibrium with them. Cells of this type, therefore, do not have completely reversible electrodes. Since thermodynamic equilibrium is not obtained in situ during the experiment, the composition of the material being studied must be established before the material is placed in the cell, in order that meaningful transference measurements may be obtained. In addition, the composition must remain invariant during the experiment. Thus care must be exercised to avoid electrolysis, or reactions between the materials being studied and the electrodes, the "protective" electrolytes, or the gaseous atmosphere. From a practical point of view this means that the transference number of the material being studied should be a function of temperature only. In a strict sense this would apply only to (a) stoichiometric compounds where the conductivities of the cations, anions or electrons were controlled by defects created by thermal disorder (e.g., Frenkel, Schottky, or intrinsic electronic disorder), or (b) compounds where the conductivity is controlled by the defects produced by the presence of impurities. Thus in practice this technique has been limited primarily to materials which exhibit only a very small departure from stoichiometry (e.g., alkali halides, alkaline earths, silver halides, lead halides, thallous halides).

The temperature range in which these compounds are measured is also very important. At low temperature the concentration of defects produced by impurities may exceed those produced from intrinsic disorder; at high temperatures intrinsic disorder predominates if the material is sufficiently pure. Thus at low temperatures both the magnitude of the conductivity and the transport numbers are very sensitive to the purity of the crystals. These two regions can generally be determined from a plot of $\log \sigma$ vs $1/T$. This type of plot is usually composed of two straight portions; the low-temperature region is generally structure sensitive and depends on the purity of the sample, while the high-temperature region is usually

EXPERIMENTAL TECHNIQUES

controlled by the intrinsic behavior of the material. Lidiard [112] has used the results of conductivity and transport study on "pure" KCl, recrystallized KCl, and KCl doped with 2×10^{-4} $CaCl_2$ by Kerkhoff [113] to illustrate this point.

Another important consideration the experimenter should be aware of when making conductivity or transference measurements is the effect of the microstructure of the specimen employed. Aside from the problem of purity, the results obtained from sintered specimens may differ from those obtained from single crystals. For example, the ionic conductivity of sintered specimens may be enhanced because of surface or grain-boundary diffusion; electronic conductivity is generally lowered for sintered specimens because of additional scattering arising from grain boundaries, pores, etc.

Since real electrodes are not completely reversible the question arises as to how closely they approximate reversible behavior. When current flows through the cell, electrons and/or ions are transported between the electrodes and the mixed conductor. Since this process requires a driving force, the electrode cannot be in exact equilibrium with the mixed conductor. This irreversible behavior causes polarization voltages to be built up in the electrode region. These voltages can affect the apparent conductivity measured.

As an example of the irreversible behavior of the cell shown in Fig. 28(b) the results of Brook et al. [114] and Yanagida et al. [115] are summarized below. The direct-current/voltage characteristics of a symmetrical cell

$$Pt, O_2(I) \mid Zr_{0.85}Ca_{0.15}O_{1.85} \mid Pt, O_2(II)$$

with $P_{O_2}(I) = P_{O_2}(II)$ was measured between 520-685°C with oxygen pressures from 1 to 10^{-20} atm using platinum electrodes of various types. For applied voltages between 0 and approximately 3 V sputtered electrodes and porous-paste electrodes fired at 800°C were found to have an ohmic I-V characteristic with a resistance equal to the bulk ionic resistance of calcia-doped zirconia. Thus the rate-controlling mechanism with electrodes of these types was ionic conduction through the zirconia. The other Pt electrodes were obtained using foil, fluxed paste, and unfluxed paste fired at higher temperatures (~1300°C). The I-V characteristics of electrodes of these types were markedly nonohmic. The rate-controlling step responsible for this behavior was confined to the cathode area. Several different rate-limiting mechanisms were proposed for the various types of platinum electrodes, depending on the value of the applied field. For example, with electrodes of Pt foil the rate-controlling processes were diffusion of oxygen through the platinum at low fields and electronic conduction through zirconia at high fields. On the basis of these studies it appears that the thinnest possible electrode (e.g., sputtered Pt electrodes or unfluxed Pt paste fired at low temperature) should be used to minimize the effects of irreversible behavior.

Although the above conclusions were based on cells of the type shown in Fig. 28(b) with $Zr_{0.85}Ca_{0.15}O_{1.85}$ as the mixed conductor and gas/inert-metal electrodes (i.e., $Pt, O_{2(g)}$) it would be expected that similar conclusions would be obtained for other mixed conductors with gas/inert-metal electrodes.

The occurrence of irreversible behavior at the electrodes can usually be detected from the current-voltage-time characteristic of the cell. Before using the above technique to measure ionic conductivity it is advisable to check the reversibility of the electrodes by investigating their I-V characteristics. As an example Heyne [116] has measured the voltage-time and the voltage-current behavior of a sample of stabilized zirconia (15 mole % CaO) with thinly sputtered Pt electrodes. These measurements were made with air at both electrodes, at 550°C. Below approximately 3 V the current is proportional to the applied voltage. At higher voltages, the current increases more rapidly with increasing applied voltage. In this region the voltage also decreases with time under constant-current conditions. These results indicate that at low voltages the electrodes are reversible. At higher voltage the bulk resistance decreases during current flow. This behavior is attributed to a loss of oxygen which results in the appearance of appreciable electronic conductance. The initial conductivity was approximately 99.99% ionic. The electronic conductivity was confirmed by a determination of the amount of oxygen transported. The amount of oxygen transported showed a corresponding decrease in time as did the resistance.

These results indicate that the sputtered electrodes act reversibly up to about 3 V for the temperature and oxygen pressure investigated. Above this voltage, the take-up of oxygen at the cathode may be rate limiting and electrolysis may result.

2. Mixed Conductor Between Different Reversible Electrodes

The ionic transference number of a mixed conductor may also be determined from emf measurements on cells with the mixed conductor between different reversible electrodes. With cells of this type the thermodynamic potential of the components of the compound are fixed at each electrode. Since each electrode has a different thermodynamic potential the composition of the compound at each electrode is different. Under open-circuit conditions the net current for this cell is zero. However, because of the gradient in chemical potentials the cations migrate toward the cathode and the anions and electrons toward the anode. Equations for the open-circuit emf of cells of this type have been derived by Wagner [117] in conjunction with the theory of oxidation of metals at elevated temperatures. The derivation of these equations has been reviewed more recently [108,109,116]. These results may be summarized as follows. If polarization is disregarded and the variability of the metal-to-nonmetal ratio in the mixed conductor $M_a X_b$ is small, the open-circuit emf E of a cell with electronic probes is given by the relation

$$E = -\frac{1}{q} \int_{\mu_M'}^{\mu_M''} t_i \frac{d\mu_M}{Z_M} = \frac{1}{2q} \int_{\mu_M'}^{\mu_M''} t_i \frac{d\mu_{X_2}}{|Z_X|} \qquad (32)$$

where t_i is the ionic transference number, q is the absolute value of electronic charge, Z_M and Z_X are the valences of the cations and anions, respectively, and μ_M and μ_X are the chemical potentials of metal M and nonmetal X per atom, and the superscripts ′ and ″ denote the values prevailing at the left-hand and right-hand side of the electrolyte, respectively.

If the chemical potential difference of M or X is small and t_i is approximately constant, the following expression may be obtained from Eq. (30)

$$E = -\frac{1}{Z_M q} \bar{t}_i (\mu_M'' - \mu_M') = \frac{1}{2Z_X q} \bar{t}_i (\mu_{X_2}'' - \mu_{X_2}') \qquad (33)$$

where $\bar{t}_i$ is the average value of t_i corresponding to the range of chemical potential between μ_{X_2}'' and μ_{X_2}'.

From an experimental point of view probably the most important consideration in the use of these cells is the selection of the appropriate electrodes. Several different types of electrodes may be used. For example, consider the following:

Type i: an inert, electronically conducting electrode in contact with the gaseous form of the nonmetallic component of the compound $M_a X_b$ (e.g., Pt,$O_{2(g)}$ | Cu_2O; C(graphite), $Br_{2(g)}$ | AgBr);

Type ii: a metallic electronically conducting electrode with the same metal, M, as the metallic component of the compound $M_a X_b$ (e.g., Ag | AgBr; Cu | Cu_2O);

Type iii: an electronically conducting electrode comprised of an equilibrium phase mixture of a metal M′ and its compound, $M_a' X_b$, where M′ is a different metal and X is the same nonmetallic component as in the compound $M_a X_b$ (e.g., Ni,NiO | Al_2O_3; Cr,Cr_2O_3 | MgO).

Although several different cells may be obtained with various combinations of the above electrodes, the most frequently used combinations are represented schematically in Figs. 31 and 32.

Cells of the type shown in Fig. 31 with both electrodes of Type iii have been used quite extensively to obtain ionic-transference measurements on metal oxides (e.g., BeO [118], MgO [118-122], Al_2O_3 [123], $Th_{0.85}Y_{0.15}O_{1.925}$ [124]. The major advantages of this

EXPERIMENTAL TECHNIQUES   125

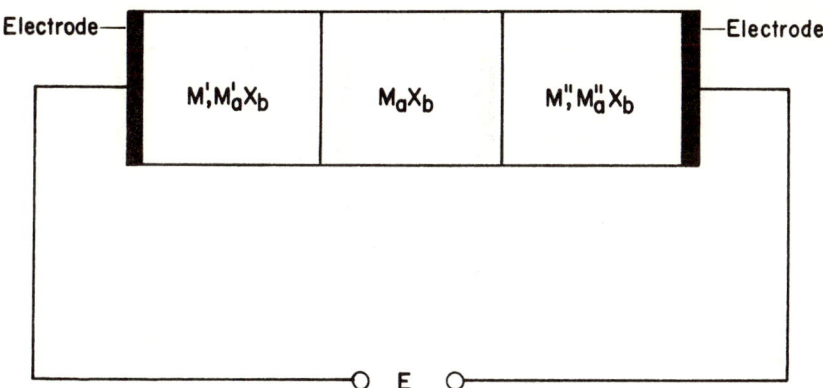

FIG. 31. Schematic diagram of a mixed conductor M_aX_b between different reversible electrodes of Type iii.

type of cell is its simplicity. For example, when BeO [118] was investigated the cell assembly consisted simply of a pellet of BeO lightly pressed between two different metal/metal-oxide electrodes, Ni-NiO and Co-CoO. The cell was then placed in an inert gas in the furnace and the emf was measured as a function of temperature. With this arrangement the electrodes which are electronic also served as the electrical contacts.

The major disadvantage of using electrodes of this type is that only a limited number of chemical potentials can be obtained and the difference in chemical potentials of the electrodes may differ significantly. Thus with this technique unless $t_i \simeq 1$ only the average value of the transference number corresponding to a range of chemical potentials can be determined using Eq. (33).

In order to determine t_i as a function of μ_M or μ_{X_2} from Eq. (33) the chemical potentials at both electrodes must be varied simultaneously and the difference in chemical potentials between the electrodes must be small. With the electrode arrangement shown in Fig. 31 this is not possible.

To circumvent this problem the cell shown in Fig. 32, which employs gas electrodes, is frequently used. Gas electrodes may be used if the μ_{X_2} in the compound M_aX_b can be controlled over a range

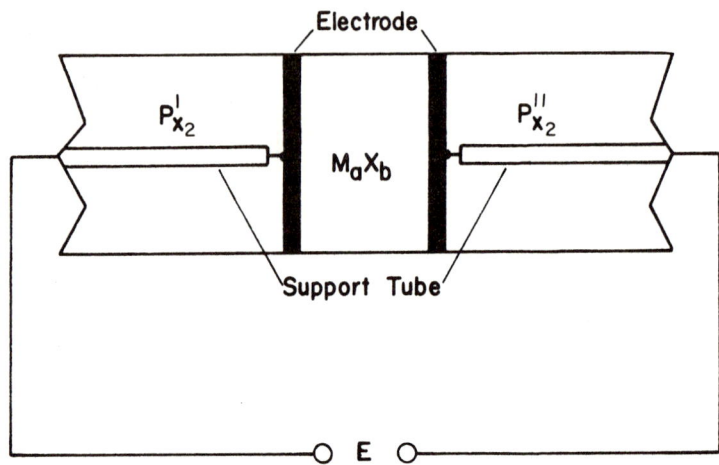

FIG. 32. Schematic diagram of a mixed conductor M_aX_b between different reversible electrodes of Type i.

of values by varying P_{X_2}. The various techniques for either direct or indirect control of P_{X_2} have been discussed above in Section III,B. Since $\mu_{X_2} = \mu_{X_2}^{\,o} + kT \ln P_{X_2}$ the emf of this cell may be represented by the expression

$$E = \bar{t}_i \frac{kT}{Zq} \ln \frac{P_{X_2}''}{P_{X_2}'} \qquad (34)$$

The major advantage of this type of cell is that in principle P_{X_2} may be varied continuously over a wide range of values. Thus if t_i is not a strongly varying function of P_{X_2} then the ratio P_{X_2}''/P_{X_2}' may be made small enough that $\bar{t}_i \simeq t_i$ and t_i may be calculated as a function of P_{X_2} from Eq. (34). The transference numbers of several metal-oxide systems (e.g., Al_2O_3 [125], CeO_{2-x} [126]) have been obtained from emf measurements on cells of this type.

One major experimental difficulty with the set-up shown in Fig. 32 is the requirement that the gaseous atmospheres at the two electrodes be completely isolated from one another. Several different techniques have been used with varying degrees of success to achieve this objective. The simplest method of separating the gas electrodes is to clamp the sample between the ends of ceramic tubes, as shown in Fig. 33. Since this arrangement is not completely gas

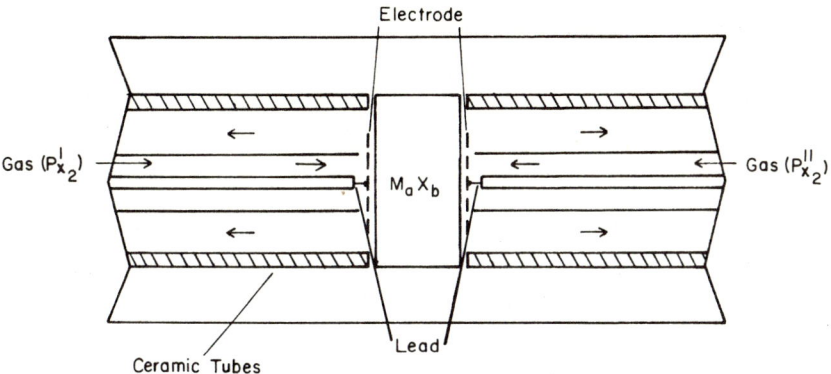

FIG. 33. Electrochemical-cell arrangement used to separate the reversible gas/inert-metal electrodes.

tight the emf frequently exhibits a flow dependence which is indicative of a leak. To reduce these leaks a platinum gasket is frequently placed between the sample and the ceramic tubes. If the system is placed under a compressive load at elevated temperatures the platinum usually forms a bond between the sample and the ceramic tubes. Platinum paste is also used to improve this bond. From a practical point of view the effect of leaks may be reduced by using a vacuum in the region between the outer ceramic tube and the ceramic tubes which are joined to the specimen. Another method that has been used to obtain a gas-tight seal is to place a pyrex ring between the sample and the ceramic tubes [126].

Another type of cell that may be used to determine t_i as a function of P_{X_2} and T is to fix P'_{X_2} and measure E as a function of P_{X_2}''; t_i can then be determined from the slope of an experimental E vs $\ln P_{X_2}''$ plot and from the expression

$$\left[\frac{dE}{d \ln P_{X_2}''}\right] = \frac{kT}{2|Z_X|q} (t_i) \quad P_{X_2} = P_{X_2}'' \tag{35}$$

Equation (35) was obtained by differentiating Eq. (32) with respect to the upper integration limit and substituting the relation $d\mu_X = (kT/2) \, d \ln P_{X_2}$.

The derivation of Eqs. (32)-(35) assumes that a well defined chemical potential or equilibrium partial pressure of M and X exists at each electrode-electrolyte interface. For an actual cell, however, these conditions may not be obtained because current-leakage paths and reactions of the gaseous environment with the electrodes and the electrolyte may disturb local thermodynamic equilibrium and produce undefined chemical potentials of the components at the electrode-electrolyte interfaces. Since a general discussion of these problems has been presented in Section III,A on galvanic-cell measurements only the major experimental problems associated with the above type of cell will be discussed in this section.

The first consideration in selecting an electrode should be its compatibility with the electrolyte being investigated. The chemical potential or partial pressure of the component fixed by the electrode should be within the stability limits of the electrolyte being investigated. For example, if TiO_{2-x} is being investigated, the P_{O_2} of the electrodes should be greater than the lower stability limits of P_{O_2}'' [127]. In addition, the mutual solubility of the electrode and the electrolyte should be negligible and no intermediate reaction products should form between the electrode and the electrolyte.

The major limitation of the use of cells of the above type is generally caused by current-leakage paths, either internal or external. The net effect of the current-leakage paths is to produce a mass transfer through the electrolyte via an ionic current. Polarization of the electrodes will occur if the ionic current is large enough to disturb local thermodynamic equilibrium and produce an undefined chemical potential of M or X at the electrode-electrolyte interfaces.

If the external leakage paths are negligible, then for open-circuit conditions the total current (i.e., the sum of the electronic current J_e and the ionic current J_i) is zero for the above type of cells. However, if $t_i < 1$ internal leakage will result from mass transport through the electrolyte via an ionic current. For this case J_i, which is equal to $-J_e$, may be calculated from the following relationship [116]:

EXPERIMENTAL TECHNIQUES

$$J_i = -J_e = \frac{\sigma_t}{2|Z_x|q} t_e t_i \text{ grad } \mu_{X_2} \qquad (36)$$

where σ_t is the total conductivity of the electrolyte. According to Eq. (36) the major factors that control the ionic current under open-circuit conditions are the values of σ_t, t_i, t_e, and grad μ_{X_2}. Polarization of the electrodes, therefore, is more likely to occur for materials that have a high conductivity and where t_i is of the same order of magnitude as t_e. When this type of electrolyte is investigated it is generally advisable to use electrodes with similar values of μ_M and μ_{X_2} so that grad μ_{X_2} is small. In this case the technique used to determine t_i as a function of P_{X_2} using Eq. (35) is generally not suitable because it requires that the grad μ_{X_2} be varied over a wide range of values. If possible the type of cell shown in Fig. 32 should be used since with gas electrodes it should be possible to make the value of grad μ_{X_2} as small as possible. It should be noted that if too small a value of grad μ_{X_2} is used, the emf of the cell will be very small and errors may arise from thermoelectric emf's. The effect of thermoelectric emf's cannot be simply canceled by reversing the gas mixtures at the two electrodes and averaging the emf's because the thermal emf's are functions of μ_{X_2}, T, and ΔT. In practice the cell is generally located in the furnace in a constant-temperature region, so any thermal emf's are essentially negligible. The average $\bar{t}_i$ is then determined for decreasing values of $\Delta\mu_{X_2}$ until either $\bar{t}_i$ is independent of $\Delta\mu_{X_2}$ or the E of the cell becomes so small that appreciable error arises in its measurement. If E does become very small a thicker sample should be used and the above procedure repeated. For fixed difference in μ_{X_2} a thicker specimen will decrease grad μ_{X_2} and thus reduce the polarization of the electrodes. With a gas-type electrode the flow dependence of the emf of the cell and the type of electrical contact should also be investigated. Van Handel [126], for example, found that the emf of the cell

$$Pt, P_{O_2}' \mid CeO_2 + CaO \mid Pt, P_{O_2}''$$

increased with increasing flow rate when platinum paste was used as the electrical contact on the sample. When the Pt lead of a Pt,Pt-Rh thermocouple was used as the electrical contact the emf was essentially independent of flow rate over a wide range of flow rates. At very high flow rates, however, cooling occurred and a thermal emf was observed.

External leakage paths are more likely to affect the transference measurements when materials with high resistances are investigated. In this case, leakage paths may arise from either surface or gas-phase conduction. A number of investigations [125,128-131] have shown that for high-resistance materials, particularly at high temperatures, the conductivity of the gas phase around the sample can be comparable to or greater than that of the sample. The gas-phase conduction involves thermionic emission from the sample, the leads, or the supporting structure and is probably dependent on the design of the experimental system. The effects of gas-phase and surface conduction may be either eliminated or minimized using a guarded-electrode technique. As an example, a schematic diagram of the guarded system used by Brook et al. [125] to measure the transference number of Al_2O_3 is shown in Fig. 34. The platinum shielding which acted as the guard was maintained at the same potential as the Pt-wire contact by adjusting the potential of the guard until no current flowed. The Keithley 600A electrometer was used as a null indicator. Contacts were formed by pressing Pt wire against the sample surface. The emf developed by the oxygen potential difference was measured with a Keithley 610B electrometer. The guard and the electrode boosted to the same potential were maintained at the same temperature and oxygen pressure to eliminate radial gradients. From measurements on the cell

$Pt,O_2 \mid Al_2O_3 \mid Air, Pt$

with a guarded circuit they observed that the measured emf agreed with the theoretical values within experimental error indicating that $t_i \simeq 1$ and that the migrating species have conventional charges. Without the guard t_i appeared to decrease with increasing temperature.

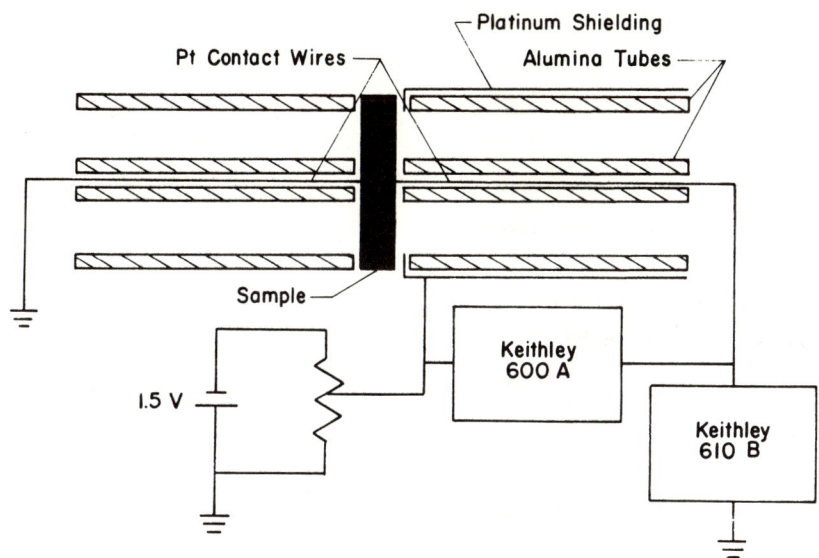

FIG. 34. Schematic of arrangement used by Brook et al. [125] to measure the emf of a cell with oxygen electrodes. The guard and the electrode, which are at the same potential, should be under identical conditions of P_{O_2} and temperature.

This reduction in the apparent value of t_i was attributed to an increase in the conduction of the gas phase.

B. POLARIZATION MEASUREMENTS

The above techniques, which use reversible electrodes, are the most frequently utilized methods for determining transference numbers; dc polarization techniques, however, can also be used to determine partial ionic and electronic conductivities. Generally cells with one reversible electrode and one electrode blocking either ions or electrons are used in the dc polarization measurements. Since most of the investigations have been made with one electrode blocking ions only this type of cell will be discussed here.

An example of a cell with a mixed conductor $M_a X_b$ and one reversible and one inert electronically conducting electrode is shown in Cell IV:

Cell IV Reversible electrode $\quad \underset{x=0}{\overset{-}{|}} M_a X_b \underset{x=L}{\overset{+}{|}}$ Inert electronically conducting electrode

Under open-circuit conditions, the chemical potential of M and X in the specimen should be fixed by the reversible electrode. However, when a potential difference is applied to the cell with the reversible electrode negative and the inert electrode positive, current flows from the inert to the reversible electrode. The M ions and holes will migrate to the reversible electrode and the X ions and electrons to the inert electrode. If the reversible electrode is completely reversible to the electronic and ionic charge carriers and the inert electrode is reversible to electron but completely blocks the passage of X ions, then the ratio a/b of M ions to X ions will increase in the left-hand side of the sample and decrease in the right-hand side of the sample. Thus under steady-state conditions only electronic current density J_e will flow because the concentration gradient of the ions will be balanced by the potential gradient and the current density J_{ions} for the ions will be zero.

Wagner [109] and Kroger [108] have derived equations for this type of cell that relate the electronic conductivity σ_e and the chemical potential μ_M of M to the applied voltage E. These equations are discussed below.

$$J_e = \sigma_e \frac{dE}{dx} \qquad (37)$$

$$\mu_{M_x} - \mu_{M_{x=0}} = Z_M q E_x \qquad (38)$$

where μ_{M_x} and $\mu_{M_{x=0}}$ are the chemical potentials of M at x = 0 and x = L, respectively, and E_x is the potential difference measured with electronically conducting probes between position x and x = 0. The use of these equations requires the use of probes to measure E_x at different locations for a current density.

EXPERIMENTAL TECHNIQUES

Since data of this type are generally inconvenient to obtain experimentally, Wagner [109] has proposed an alternative method by which the electronic conductivity can also be deduced from the current density as a function of potential E, between $x = 0$ and $x = L$, if the potential drop in the current leads and contact resistance at the phase boundaries can be disregarded. The equations he obtains are

$$(\sigma_e)_{x=L} = L \left[\frac{dJ_e}{dE}\right]_{x=L} \tag{39}$$

and

$$\mu_{M_{x=L}} = \mu_{M_{x=0}} - Z_M q E_{x=L} \tag{40}$$

With these equations the electronic conductivity of a compound $M_a X_b$ may be obtained as a function of μ_M by measuring the resulting current density for different applied potentials $E_{x=L}$.

In most of its applications Cell IV has been used to study electronic conduction in the halides of Ag [132], Pb [133], and Cu [134]. In these cells the reversible electrode is the metallic electrode and the inert electronically conducting electrode is usually Pt or graphite (e.g., Cu|CuX|graphite, where X denotes Cl, Br, or I). The current-voltage (I-V) relationships may be obtained by either applying a constant current and measuring the resulting voltage after steady-state conditions are obtained or by applying a fixed potential and measuring the resultant current. Both methods should give consistent results. The data for these cells have been analyzed using the following equation derived by Wagner [109]

$$J_e = \frac{I}{A} = \frac{kT}{qL} \{\sigma_e^\circ [1 - \exp(-qE/kT)] + \sigma_h^\circ [\exp(qE/kT) - 1]\} \tag{41}$$

where A is the cross-sectional area, L is the thickness of the sample, q is the charge on an electron, σ_e° and σ_h° are the electronic conductivities in the sample $M_a X_b$ coexisting with the metallic electrodes due to excess electrons e and electron holes h, and the other symbols have their usual meaning.

In the derivation of Eq. (41) [108,109] it is assumed that excess electrons and holes follow the laws of ideal dilute solutions, that their mobilities are independent of concentration, that the change in the concentration of atomic defects arising from thermal disorder with variation in the ratio a/b is small, and that decomposition of M_aX_b is negligible.

The variation of J_e with E that is observed depends on the values of σ_e^o and σ_h^o. If $\sigma_e^o > \sigma_h^o$, according to Eq. (41) J_e should increase at a decreasing rate with increasing E until a plateau is reached where

$$J_e \simeq \sigma_e^o \frac{kT}{qL} \qquad (42)$$

Ilschner [132] has observed this plateau for AgBr. To illustrate this Ilschner's results are shown in Fig. 35. The temperature dependence of the electronic conductivity of AgBr in equilibrium with pure silver was then determined from the plateau shown in Fig. 35 and from Eq. (42).

At higher applied voltages the hole conductivity becomes more important and hence the second term in Eq. (41) can no longer be neglected. Equation (41) may be approximated by the following relation:

$$\log J_e = \log \frac{1}{A} = \log \left[\frac{\sigma_h^o kT}{qL}\right] + \frac{EF}{2.303kT} \qquad (43)$$

if $E \gg kT/q$ and $E \gg [kT/q] \ln (\sigma_e^o/\sigma_h^o)$.

The electron-hole conductivity of cuprous halides was deduced by Wagner and Wagner from current-density vs potential curves for the cell

Cell V   Cu | CuX | graphite

where X denotes Cl, Br, or I [134]. Isothermal plots of log J vs E

EXPERIMENTAL TECHNIQUES   135

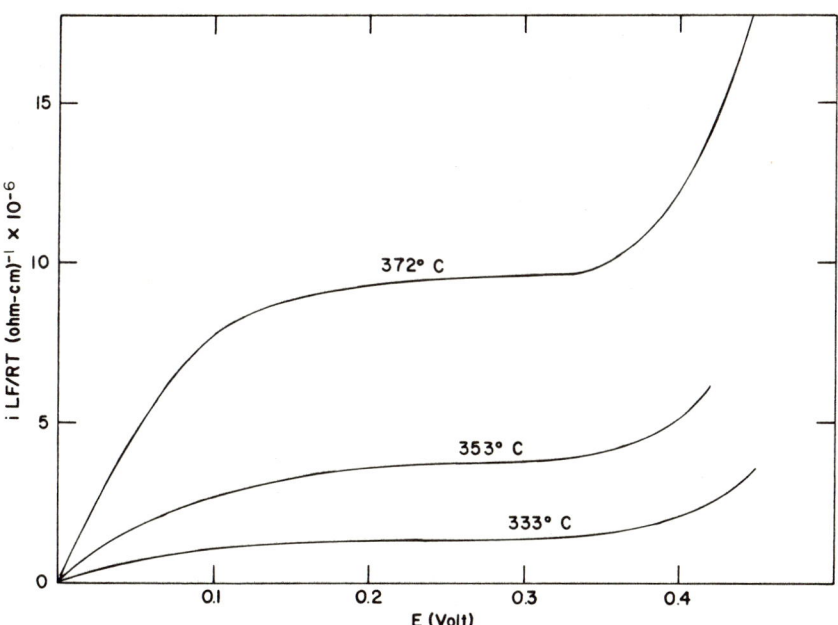

FIG. 35. Isothermal current-potential behavior of AgBr reported by Ilschner [132].

for Cell V yielded straight lines whose slopes were close to the value of $q/2.303kT$ suggested by Eq. (43). Values of σ_h° were then determined from Eq. (43) for several different temperatures in the temperature range 250-450°C. By comparing the ratio of total conductivity to hole conductivity Wagner and Wagner observed that the transference number for holes in CuCl, CuBr, and CuI was very small [134]. For example, the transference number of holes in CuBr equilibrated with copper at 400°C was 2×10^{-5}. The low-temperature limit, approximately 250°C was determined by the lowest temperature at which reproducible data could be obtained. The voltage E of the cell was varied between approximately 0.2 and 0.7 V. At lower voltages, where the current is very small (<10 μA), hysteresis was observed. The nature of these irregularities was not discussed. The problem was avoided by operating at voltages greater than 0.2 V. According to Eq. (41) the current density should be inversely proportional to the thickness L of the sample. This relation has been found to hold for samples with $L > 0.25$ cm. Deviations found for smaller values of L were ascribed to contact resistance between the CuX sample and the electrodes.

Ilschner [132], using a cell of the type

Cell VI   Ag | AgX | graphite

also observed hole conduction in the silver halides AgI and AgCl. With a AgI sample, log I vs E exhibited a linear dependence as predicted by Eq. (43); the slope, however, was equal to $1.2q/kT$ instead of the predicted value, q/kT. The cell exhibited a current-voltage hysteresis which tended to limit the minimum voltage employed to 200 mV. The maximum voltage was limited to 400 mV because above this level voltage decomposition became important. When AgCl was investigated the current exhibited an exponential dependence on E. The slope, however, was equal to approximately $0.3q/kT$ instead of the predicted value, q/kT. No explanation for this behavior was given. In addition the reproducibility of this cell was poor; in particular, the scatter with respect to temperature dependence was large. These inconsistencies were attributed to some process other than electronic conduction: possibly the migration of impurities. Because of these problems, only qualitative results were presented for this cell.

In addition to these problems, the above analysis of the data obtained from Cells V and VI provided information about the electronic conductivity at only one composition (i.e., the composition that coexists in equilibrium with the metal electrode). In a more recent dc polarization study Patterson et al. [135] have determined the electron and hole conductivity of both $Zr_{0.85}Ca_{0.15}O_{1.85}$ and $Th_{0.85}Y_{0.15}O_{1.925}$ as a function of P_{O_2}. They used cells of the types shown below, with one electrode reversible to oxygen and one electrode reversible to electrons only.

Cell VII M, MO_x or $M'O_x, M'O_y$ | $Zr_{0.85}Ca_{0.15}O_{1.85}$ | Au or Pt

EXPERIMENTAL TECHNIQUES 137

Cell VIII $\quad\begin{array}{|c|}\text{M, MO}_x \\ \text{or} \\ \text{M}'\text{O}_x, \text{M}'\text{O}_y\end{array}\Bigg|\ \text{Th}_{0.85}\text{Y}_{0.15}\text{O}_{1.925}\ \Bigg|\ \text{Au or Pt}$

where M = Cu, Ni, CO, Fe, Cr, Nb, V, Ti, and M' = Cu. They carried out an extensive series of preliminary experiments to determine the effect of various experimental conditions on the steady-state current. For example, the steady-state current was established for a given emf and temperature while the He flow rate, total gas pressure, Pyrex wall temperature and cold-trap levels were varied. Because all these changes affected the steady-state current the system was redesigned so that all greased joints were positioned on the upstream side of the liquid-nitrogen cold trap and all the tubing between the cold trap and the furnace chamber was maintained at 200-300°C.

In order to further minimize any reactions between specimen and surrounding gas purified He was set at a P_{O_2} approximately that of the cell by passing high-purity He over a boat immediately upstream from the cell, filled with the two-phase electrode mixture of the cell. The net effect of these procedures was to produce a stable steady-state current by minimizing the exchange of oxygen between the sample and its gaseous atmosphere.

Isothermal measurements of J vs E were made for Cells VII and VIII for each type of reversible electrode at 800°C, 900°C, and 1000°C. The previous investigations [132,134,136,137] involving this type of dc polarization technique were restricted to experimental conditions where either σ_e^o or σ_h^o predominated. The limiting-case analysis was then made using either the current plateau of J vs E for determining σ_e^o [Eq. (42)] or the intercept of log J_e vs E [Eq. (43)] for determining σ_h^o. Patterson et al. [135] used a different method of analyzing the data, a method which is valid in the above limiting-case situation as well as when σ_e^o and σ_h^o are comparable in value. The expression they used,

$$\frac{J_e}{1 - \exp(-qE/kT)} = \frac{kT}{qL}\left[\sigma_e^o + \sigma_h^o \exp(qE/kT)\right] \qquad (44)$$

was obtained by dividing both sides of Eq. (41) by $[1 - \exp^{(-qE/kT)}]$. If the data fit the above model then a plot of $J_e/[1 - \exp^{(-qE/kT)}]$ vs $\exp(qE/kT)$ should give a straight line of slope $(kT/qL)\sigma_h^\circ$ and intercept $(kT/qL)\sigma_e^\circ$. The data they obtained were converted to $J_e/[1 - \exp^{(-qE/kT)}]$ and $\exp(qE/kT)$ values and plotted to determine the values of σ_e° and σ_h° corresponding to equilibration of the specimen with the μ_{O_2} fixed by the reversible electrode. A comparison of the P_{O_2} dependence of the electron, hole, and total conductivities for the $Zr_{0.85}Ca_{0.15}O_{1.85}$ and $Th_{0.85}Y_{0.15}O_{1.925}$ electrolytes indicated that ionic conductivity predominated over the entire P_{O_2} region investigated and that $\sigma_h > \sigma_e$ in the high-P_{O_2} region.

Although this polarization technique appears to be generally applicable, it should be noted that experimental difficulties were encountered in the application of this technique at very low oxygen pressures and in the determination of the conductivity of the minority electronic carrier. For example, for Cell VII at high P_{O_2} the value of σ_e did not appear to be significant because of background currents, and at very low P_{O_2} the cell exhibited ohmic behavior as well as other inconsistencies.

In the investigation of Cell VIII another problem was encountered: σ_e was very small and appeared to be independent of P_{O_2}, which indicated that the actual value of σ_e in $Th_{0.85}Y_{0.15}O_{1.925}$ was below the limits of resolution for the apparatus employed. Patterson [135] attributed this problem to leakage conductivities around the specimen that exceeded the σ_e through the bulk of the specimen. Electron transport through the gas phase was suspected as the principal leakage because preliminary experiments in which a void space (He gas) replaced the specimen yielded the same magnitude for this leakage conductance. In addition it was observed that the leakage conductance showed a positive temperature dependence and that it increased when the system was evacuated. Patterson et al. [135] suggest that the principal electronic leakage is probably a result of thermionic emission and that this leakage could be considerably

EXPERIMENTAL TECHNIQUES   139

reduced for polarization studies by the placement around the cell of
a metallic cylinder which would be maintained at a high positive
potential relative to the cell.

Polarization measurements have also been made on electrolytes
of $Zr_{0.85}Ca_{0.15}O_{1.85}$ by Vest and Tallan [139] and on $0.13YO_{1.5}$-
$0.87ThO_2$ by Wimmer et al. [140]. The polarization technique used by
these investigators consists of measuring the three-terminal conductivity with relatively-high-frequency ac and dc. A brief description of this technique as presented by Wimmer et al. [140] is given
below. The equivalent circuit assumed to represent the specimen is
shown in Fig. 36, where G_e is the electronic conductance, G_t is the
ionic conductance, C represents the blocking condition for ions due
to the platinum electrodes, and G_{e-g} is the electrode-gas conductance.
Analyzing the response of this circuit at zero and infinite time
results in a zero-time conductance of the parallel combination of G_e
and G_t and an infinite-time conductance of G_e if the electrode-gas
conductance is small compared to the electronic conductance. If G_{e-g}
is comparatively large, however, the steady-state dc conductance can
be controlled by ionic conduction through the specimen, with oxygen
being absorbed from the gas phase at one electrode and discharged at
the other electrode. In the absence of electrode-gas conductance,
the transference numbers for ions and electrons can be calculated
from the relative values of the total (zero-time) and electronic
(infinite-time) conductances.

One of the major criticisms of this polarization technique is
that since the mixed conductor is placed between electronic conductors the initial metal-to-nonmetal ratio of the sample must be
well defined. For example, the polarization study of ThO_2 by Danforth and Bodine [141] using a cell of the type

Cell IX   Pt | ThO_2 | Pt

in a vacuum is unsuitable because the P_{O_2} is ill defined and thus
the composition of the compound is also ill defined. Vest and Tallan
[139] attempted to circumvent this problem by first equilibrating the

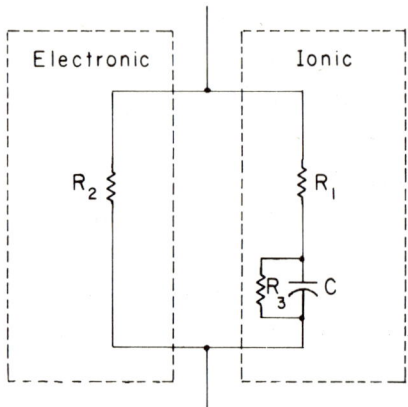

FIG. 36. Equivalent circuit of a mixed conductor with electrodes blocking ions only, according to Vest and Tallan [139]. $R_2 = 1/G_e$; G_e = electronic conductance, $R_1 = 1/G_i$; G_i = ionic conductance; $R_3 = 1/G_{e-g}$; and G_{e-g} = electrode-gas conductance.

oxide specimen with a given P_{O_2} in the atmosphere to fix the initial oxygen content. However, with this method, in order to obtain meaningful results no oxygen loss or gain should occur during the establishment of the stationary polarized state. In practice it is likely that some change in composition occurs. The reason for this is that there is no reversible electrode present, as used in the above polarization techniques, that will define the composition and the chemical potentials in the sample adjacent to the reversible electrode. Another problem with the technique used by Vest and Tallan [139] is the assumption that the electronic and ionic conductivities are independent of one another (i.e., the blocking of the ionic current does not affect the electronic conductivity). This is a limiting-case assumption applicable only for small applied voltages, corresponding to the small differences in μ_{O_2} (i.e., the variation in composition and μ_{O_2} across the sample is small enough that the

EXPERIMENTAL TECHNIQUES 141

electronic and ionic conductivities may be represented by average values).

Since several of the experimental problems associated with the polarization measurements have been discussed throughout this section, the discussion presented here will be limited primarily to the experimental conditions assumed in the derivation of Eqs. (37)-(43).

1. The Assumption of Negligible Electrolytic Current, $J_i \simeq 0$

Decomposition of the sample will produce ionic currents. To avoid this the maximum potential difference applied to the sample is usually kept small to avoid direct decomposition of the specimen at the electrodes. Decomposition, however, may also occur by a reaction between the sample and gaseous atmosphere around the sample. For example, under an applied field the ratio a/b of the compound $M_a X_b$ will decrease as one crosses the sample from the negative electrode to the positive electrodes. The chemical potential of X, therefore, will be higher near the positive electrode, which will result in a higher vapor pressure and thus a greater tendency to vaporize the nonmetallic component. Wagner and Wagner [134] have calculated the upper limit of current for decomposition of CuCl and CuBr by determining the equilibrium partial pressures of halogen atoms and molecules, P_X and P_{X_2} at the graphite electrode (Cell V) for a given value of the applied voltage E and the expression

$$I_{decomp} = I_X + I_{X_2} = Fn\left[\frac{P_X}{P}\right] + 2Fn\left[\frac{P_{X_2}}{P}\right] \qquad (45)$$

In Eq. (45), I_{decomp} is the decomposition current, I_X and I_{X_2} are the partial currents resulting from evolution of halogen as atoms and molecules, respectively, n is the flow rate of argon in moles/sec flowing over the cell, and P is the total pressure.

Ionic currents may also occur because of incomplete blocking of ions at the inert electronically conducting electrode. This problem usually arises from an electrode-gas interaction. For example, when polarization studies are made on metal oxides particular care must be exercised to avoid reactions of the type

$$O^{2-}_{(surface)} \rightarrow \tfrac{1}{2}O_{2(g)} + 2e_{(electrode)}$$
$$\tfrac{1}{2}O_{2(g)} + 2e_{(electrode)} \rightarrow O^{2-}_{(surface)}$$

between the specimen, the electrode, and the oxygen in the surrounding atmosphere. The type of electrode and the method of application are very important in determining the behavior of the electrode. Several investigators have studied the effects of different types of electrodes and methods of application on the current-time and current-voltage characteristics on ZrO_2-CaO specimens. In general the degree of blocking was observed to increase with increasing thickness of the electrode. Thick sputtered platinum or platinum paste showed more blocking action. The degree of blocking was also shown to be a function of P_{O_2}, current, and time.

Since there is no straightforward experimental test to determine if $J_i \simeq 0$, it is usually advisable to see if consistent results can be obtained when different blocking electrodes and a range of flow rates for the gaseous atmosphere are used.

2. The Assumption that a Well-Defined Composition and Chemical Potential Gradient Exists Across the Sample

As discussed above, a suitable reversible electrode is usually selected to insure that the composition and the chemical potential of the components in the sample in contact with the electrode are well defined. The problems associated with reversible electrodes are not discussed here, since they have already been discussed in detail in Section III,A.

In principle the composition and the chemical potential of M and X are fixed at one end of the sample by the reversible electrode and at the blocking electrode by the magnitude of the applied potential. The derivation of Eqs. (37)-(43) are based on a gradient in one dimension only. The uniformity of the potential gradient inside

EXPERIMENTAL TECHNIQUES   143

the sample, however, may be upset by an interaction between the sample
and the surrounding gas. For example, if P_{X_2} of the sample is high
and an inert gas flows over the sample, decomposition may occur as
discussed above. On the other hand, if P_{X_2} of the sample is low, but
significantly less than P_{X_2} of the gaseous environment, the sample
will "getter" X_2 from the gas, and this will alter the composition
of the sample. To eliminate this problem a high-purity inert gas is
generally employed. When a metal oxide is investigated at low oxygen
pressures a purified inert gas may not be adequate [135]. This is
because the gas probably contains oxygen-producing impurities such as
H_2O or CO_2. In this case, in order to minimize interaction with the
surrounding gas it may be necessary to set the P_{O_2} of the gas at
approximately the same value as the P_{O_2} of the sample. Patterson et
al. [135] accomplished this by passing the gas over a combustion boat
filled with a two-phase metal/metal-oxide mixture (i.e., the same
phase mixture as used for the reversible electrode) and located
immediately upstream from the cell.

In addition to the above assumptions it should be noted that
according to Wagner [109] Eqs. (37)-(43) were derived assuming that
small changes in the ratio a/b do not affect the chemical potential
of the cations. Kroger [108] has rederived these equations assuming
the transport of atomic and electronic defects instead of only ions
and electrons as used in Wagner's treatment. In the analysis used
by Kroger the charged imperfection which dominates the electroneutrality equation determines the form of the exponential terms in
Eqs. (41)-(43).

VI. THERMOELECTRIC MEASUREMENTS

When a conductive material is placed between electrodes at
different temperatures, a potential difference $\Delta\varphi$ is produced. The
potential difference between the electrodes is the result of three
individual thermal potentials: a homogeneous potential difference

$\Delta\varphi_{hom}$ due to the temperature gradient across the specimen, a heterogeneous potential $\Delta\varphi_{het}$ due to the difference in contact potentials at the two electrode-specimen interfaces at the two temperatures, and the homogeneous potential differences $\Delta\varphi_n$ due to temperature gradients in the contact wires. The potential difference per degree temperature is termed the thermoelectric power or the Seebeck coefficient, which is given by the following expression:

$$\alpha_T = -\frac{\Delta\varphi}{\Delta T} = -\frac{\Delta\varphi_{hom}}{\Delta T} - \frac{\Delta\varphi_{het}}{\Delta T} - \frac{\Delta\varphi_n}{\Delta T} \tag{46}$$

Measurement of the thermoelectric power can provide useful information on the enthalpies of formation of point defects and their heats of transport for ionic and electronic conductors [142,143]. In order to obtain this type of information the appropriately derived expression for the terms on the right-hand side of Eq. (46) must be used. The third term on the right-hand side of Eq. (46) is usually smaller than the first two terms and is therefore negligible in many analyses [144]. If metallic leads are used where the absolute thermoelectric power is known as a function of temperature [145] (e.g., Pt, Au) then the contribution to α_T from the leads may be calculated. To avoid possible errors in the analysis it is generally advisable to calculate the contribution of this term. The derivation of expressions for the first two terms in Eq. (46), which are frequently referred to as α_{hom} and α_{het}, depends upon the nature of the predominant charge carriers in the conductor and on the type of electrodes used [146]. The derivation of these equations has been discussed for cases involving ionic conductors with reversible electrodes and for electronic conductors with inert electronically conducting electrodes [146].

Nonisothermal cells with ionic or mixed conductors and identical reversible electrodes are usually referred to as thermogalvanic cells or thermocells. These thermocells may be represented by the following types of cells:

EXPERIMENTAL TECHNIQUES

Cell X $\qquad$ $M \mid M_aX_b \mid M$
$\qquad\qquad\qquad$ T $\quad$ T+ΔT

Cell XI $\qquad$ $M', P_{X_2} \mid M_aX_b \mid P_{X_2}, M'$
$\qquad\qquad\qquad$ T $\quad\;\;$ T+ΔT

In Cell X, M is assumed to be a reversible metal electrode, and in Cell XI, M' is an inert electronically conducting electrode in contact with a gas containing a controlled partial pressure of either M or X_2. The problem associated with reversible electrodes has been described previously, in Section III,A on galvanic cells.

Cells X and XI have been used extensively to investigate the thermoelectric power in alkali halides, in silver and cuprous salts, and in their mixed crystals. More recently Ruka et al. [147] and Tallan and Bransky [148] have used Cell XI to investigate the ionic Seebeck coefficient of heavily doped metal-oxide electrolytes containing very large charge-carrier concentrations. As an illustration of the effect of the type of reversible electrodes used with an ionic conductor, consider the following cells:

Cell XII $\qquad$ Ag $\mid \alpha$ - AgI $\mid$ Ag
$\qquad\qquad\qquad$ T $\qquad\;\;$ T+ΔT

Cell XIII $\qquad$ Pt, $I_{2(g)} \mid \alpha$ - AgI $\mid I_{2(g)}$, Pt
$\qquad\qquad\qquad$ T $\qquad\qquad\;\;$ T+ΔT

where T + ΔT > T. In Cell XII the polarity of the cold electrode is positive and in XIII it is negative [149]. Since according to convention the sign of the Seebeck coefficient is determined from the sign of the cold electrode, the Seebeck coefficient of Cell XII is positive, and that of Cell XIII is negative. Although both cells use the same electrolyte the reactions at the electrodes are different. At the cold electrode, which is the cathode in Cell XII, the reaction

$$Ag \rightarrow Ag^+ + e^-$$

occurs, whereas in Cell XIII this electrode is the anode and the reaction

$$I^- \rightarrow \tfrac{1}{2}I_{2(g)} + e^-$$

occurs. The cells differ in that the carrier ions (Ag^+) move in opposite directions with respect to the temperature gradient. Hence the net transport of material, iodine in Cell XIII and silver in Cell XII, is in opposite directions.

It should be noted that for thermocells of type XI the Seebeck coefficient may be more dependent on the heterogeneous reaction occurring at the electrode-electrolyte interface than on the homogeneous portion of the thermoelectric power. For example, Seebeck coefficients for cells of the types

Cell XIV $\qquad O_{2(g)}, Pt \mid (ZrO_2)_{0.85}(CaO)_{0.15} \mid Pt, O_{2(g)}$
$\qquad\qquad\qquad\quad T \qquad\qquad\qquad\qquad\qquad T+\Delta T$

and

Cell XV $\qquad\qquad O_{2(g)}, Pt \mid ThO_2 \mid Pt, O_{2(g)}$
$\qquad\qquad\qquad\qquad\quad T \qquad T+\Delta T$

investigated by Ruka et al. [147] and by Tallan and Bransky [148], respectively, were shown to be functions of the temperature dependence of the chemical potential of oxygen, $d\mu_{O_2}/dT$, rather than the partial pressure of oxygen in the region where ionic conduction predominated. They have shown that for Cells XIV and XV the Seebeck coefficient difference for two different gas mixtures may be represented by the expression

$$(\alpha_i)_{gas\ 1} - (\alpha_i)_{ref} = \frac{1}{4F}\{[\frac{d\mu(O_2)}{dT}]_{gas\ 1} - [\frac{d\mu(O_2)}{dT}]_{ref}\} \quad (47)$$

where F is the Faraday constant. To illustrate this the results of Tallan and Bransky [148] on Cell XV are shown in Fig. 37. It should be noted that the discontinuities in magnitude and slope between segments of the α_i vs P_{O_2} plots obtained in different gas mixtures are absent in α_i vs $\Delta[d\mu_{O_2}/dT]$ plots, where the slope was equal to $1/4F$ in accordance with Eq. [47].

In contrast to the extensive measurements on cells of type X with reversible electrodes in which the electrode M has a cation in

EXPERIMENTAL TECHNIQUES   147

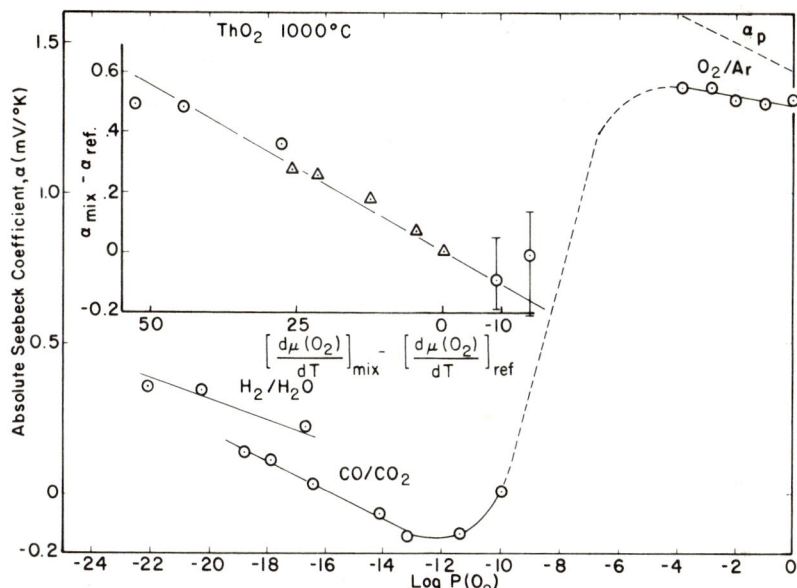

FIG. 37. Absolute Seebeck coefficient of ThO_2 as a function of oxygen partial pressure in several different gas mixtures at $1000°C$. The inset is a plot of change in Seebeck coefficient with $d\mu(O_2)/dT$ relative to an arbitrary reference value. The dashed line, α_p, is the calculated electron-hole Seebeck coefficient.

common with the salt M_aX_b, there have been relatively few investigations of the Seebeck coefficient on cells of the type

Cell XVI $M'\ |\ M_aX_b\ |\ M'$

in which the metal electrodes M' do not have a cation in common with the ionic conductor M_aX_b. For example, Seebeck coefficient measurements employing platinum electrodes have been made by Thiel [150] on PbI_2, $PbCl_2$, $TlCl$, $TlBr$, TlI, and $NaNO_3$, and by Nikitinskaya and Murin [151] on NaCl and KCl. Unfortunately, these measurements usually indicated either a high degree of irreproducibility or a thermopotential which initially fluctuated wildly and even changed sign before approaching a steady-state value. More recently, Jacobs and Maycock [152] have measured the thermoelectric power of single crystals of KCl and KCl + $SrCl_2$ between platinum electrodes over a wide range of temperatures. Unlike the analysis of Cells X-XV, their interpretation of the data is based on the assumption that the charge carriers do not cross the electrode/salt interface, which results in a different type of expression for the heterogeneous thermopotential. Thus it would be expected that the magnitude of the Seebeck coefficient for cells of type XVI would differ from the value measured with a cell that had reversible-type electrodes.

The above discussion of Cells X-XVI has been limited primarily to thermocells (i.e., ionic or mixed conductor). For electronic conductors (n- or p-type) electronically conducting electrodes are used. These are generally metallic electrodes, as shown in Cell XVI.

The above discussion of cells shows how both the magnitude and the sign of the Seebeck coefficient may be influenced, through the heterogeneous portion of the coefficient, by the type of electrode and atmosphere used. When the Seebeck coefficient of a material (e.g., M_aX_b) is investigated it is important that the appropriate type of electrode and surrounding atmosphere be selected. If M_aX_b is an ionic conductor, reversible electrodes are generally used (e.g., Cells X-XV). Reversible electrodes will not be discussed here, since they have been described previously in Section III,A on galvanic cells. Metallic electrodes (e.g., Pt, Au) are generally used for electronic conductors. A discussion of techniques for applying metallic electrodes to ceramic materials is presented in Section II,C.

Control of the ambient atmosphere is important not only because of its possible effect on the reaction at the electrodes but also in controlling the composition of the sample, particularly at elevated temperatures. The gas-equilibration techniques for controlling the composition of the specimen are discussed in Section III,B,1.

A number of different systems have been used to measure the Seebeck coefficient [148,153-155] over a wide range of temperatures. For example, Thurber and Mante [153] describe an apparatus they used to measure the Seebeck coefficient of rutile from $2°K$ to $300°K$. An apparatus for measuring the thermal diffusivity, electrical conductivity, and Seebeck coefficient from room temperature to 900°C was described by Meddins and Parrott [156]. Typical sample holders that have been used at elevated temperatures are shown in Figs. 38 and 39. Tallan and Bransky used the sample holder in Fig. 38 for thermal emf and electrical-conductivity measurements on ThO_2 up to 1600°C in a controlled atmosphere of oxygen-argon, CO_2-CO or H_2O-H_2 [148]. The sample for thermoelectric measurements was suspended from two Pt/Pt-10%Rh thermocouples whose junctions were either embedded in small conical holes drilled through the ends of the bar-shaped sample or simply wrapped around the specimen ends. A cold finger immediately below the sample was used to impose various temperature gradients (usually less than 10°C) on the specimen. The thermal emf's were measured using the Pt leads of the thermocouple.

The apparatus shown in Fig. 39 was used by Ruka et al. [147] to measure the Seebeck coefficient of a 0.85 ZrO_2-0.15 CaO electrolyte thermocell. The specimen was situated between the polished faces of two platinum blocks. The use of the platinum blocks served to minimize temperature gradients in the radial direction. On the upper block a close-fitting aluminum cylinder acted as a compression weight and in addition carried a heater winding for controlling the temperature gradient of the specimen. The lower block was supported by a calcia-stabilized zirconia conical pivot which insured that the specimen and blocks were aligned for uniform contact pressure at all times.

According to Ruka et al. the specimen's thermal emf was measured by means of 0.020-in. Pt wire attached to the Pt blocks. To measure the temperature of each block at a point as near to the specimen as possible, 0.020-in. Pt-10%Rh wire was led into the block through alumina insulating tubing. The end of this wire was then bonded to the block by means of two small Pt plugs. This produced an internal junction near the interface, whose lead wires were the Pt-10%Rh wire and the previously mentioned Pt wire.

The procedure generally used to determine the Seebeck coefficient is to measure the thermal emf E across the common leads of the

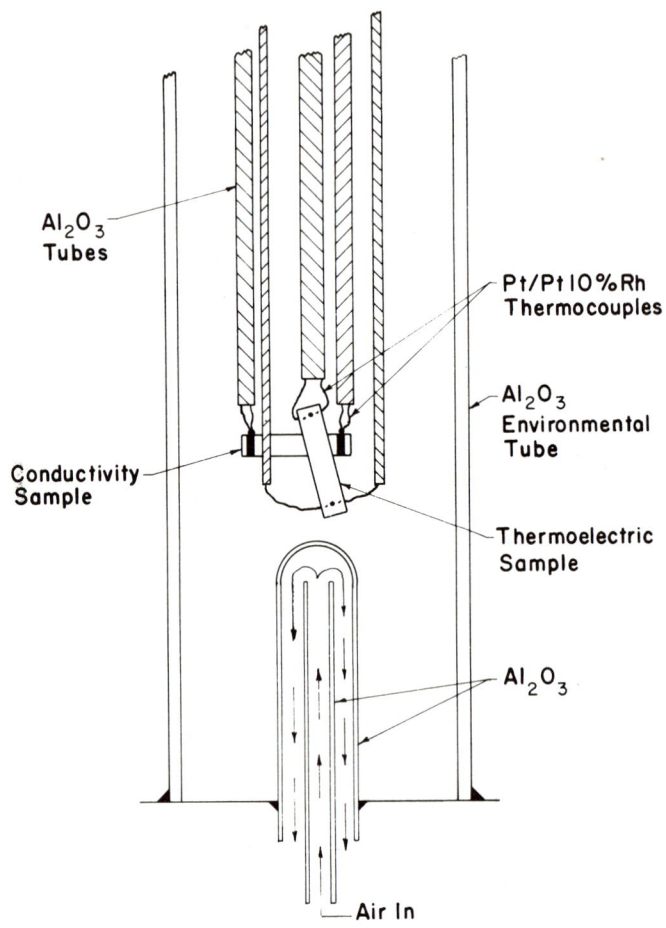

FIG. 38. Sample holder for thermal emf and electrical conductivity measurements according to Tallan and Bransky [148].

thermocouple as a function of ΔT. To avoid errors in determining the gradient in temperature ΔT it would be advisable to use calibrated thermocouples. The Seebeck coefficient α is then calculated from the slope $\alpha = |dE/d(\Delta T)|$ of a thermal emf/temperature-gradient plot. To obtain an accurate value of α, it is generally preferable to use a least-squares fit to approximately twenty or more points, preferably taken with the temperature gradient both increasing and decreasing. To avoid errors arising from the temperature dependence

EXPERIMENTAL TECHNIQUES   151

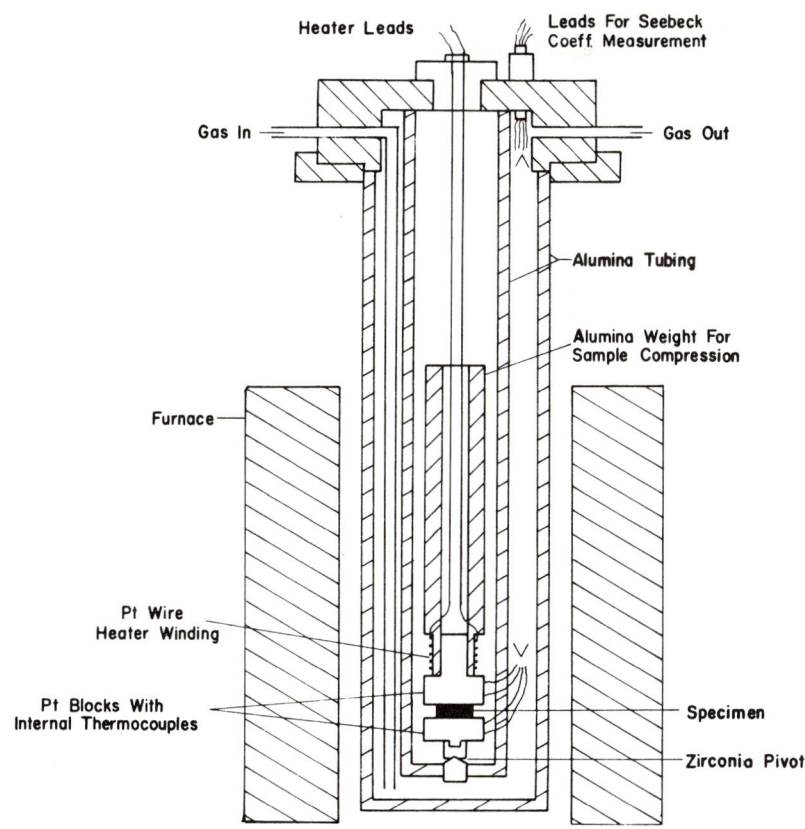

FIG. 39. Schematic of apparatus used by Ruka et al. [147] for high-temperature Seebeck coefficient measurements.

of α the maximum value of $|\Delta T|$ should not exceed approximately 10°C. The sign of α may be determined from the polarity of the cold end of the sample.

Although the measurement of the Seebeck coefficients is simple in principle, a number of experimental problems have frequently been encountered. For example, Jacbos and Maycock [152] reported that measurements on pure KCl after only a moderate annealing process exhibited an emf when ΔT was zero. They eliminated this problem by annealing the sample at 600-660°C for 48 h prior to making any measurements. They associated this problem with an uneven distribution of impurities, which was eliminated after the above annealing process.

Tallan [157] reports that with different sintered specimens of the same metal oxides parallel displacements of the plots of the Seebeck coefficient vs 1/T are frequently observed (repeatability in activation energy but not in magnitude of the Seebeck coefficient). They believe that this problem is associated with microstructural effects, perhaps microcracking. The problem can generally be minimized, at least its symptoms seem to disappear, if the sample is heated for several hours in the thermal emf sample holder to the maximum temperature of interest before beginning the temperature dependence study.

Another problem that may occur with high-resistance samples is the occurrence of leakage paths. For electronic conductors the sign of the Seebeck coefficient is determined by the carrier using the most conducting paths. For an ionic conductor this is also true for the homogeneous part of the Seebeck coefficient. Thus if external leakage paths are large the surface or gas carrier may determine the magnitude and sign of the Seebeck coefficient. For example, Brook et al. [158] have reviewed the previous electrical-conductivity, transference-number, and thermoelectric-power data on Al_2O_3 and found the data obtained by different investigators to be extremely diverse. They were able to avoid this problem associated with external leakage paths by making measurements on Al_2O_3 with a volume guard to avoid surface and gas conduction. The general technique for guarding systems has been discussed in Section IV.

VII. MOBILITY

The mobility of the charge-carrying species is often of interest when considering the electrical behavior of ceramic materials. The mobility μ of a charge carrier is its velocity per unit electric field, which is a material parameter. As defined this is often referred to as the true or microscopic mobility of a charge carrier in a material. Other terms which are related to the method of measurement of mobility are also used to designate the mobility of a charge-carrying species [159].

The Hall mobility μ_H, which is the product of the Hall constant R_H and conductivity σ, is used to designate the mobility obtained via the Lorentz force equation. The Hall mobility measurement only measures the mobility of moving charge carriers and, therefore, will not be affected by traps, as are some measurement techniques that will be discussed subsequently. The distribution of velocities of charge carriers, which is dependent on the scattering mechanism and degree of degeneracy, influences the relationship between the Hall mobility and the microscopic mobility; i.e., the microscopic mobility would be obtained by multiplying the Hall mobility by some appropriate constant r.

The conductivity mobility μ_c is calculated from the conductivity using the relationship

$$\mu_c = \frac{\sigma}{ne} \qquad (48)$$

where some other measurement has been used to obtain the carrier concentration n. The value of n may be determined by the use of appropriate impurity doping to yield carrier saturation or by interpreting thermogravimetric data. If the correct models have been used to determine the conductivity mobility, then this property is essentially identical to the microscopic mobility.

The drift mobility μ_d is the average velocity of drift per unit electric field for a charge carrier moving in an electric field. This is the same as the microscopic mobility if no charge-carrier trapping is present. If a charge carrier is trapped a portion of the time during its transit, then the measured time of drift between two points in space would be larger than the actual drift time. The calculated mobility would then be less than the true mobility by a factor equal to the ratio of the actual time of drift divided by the total measured drift time.

The three aforementioned mobilities are of most practical interest when considering ceramic materials. Other terms are used to designate mobility where the charge-carrier mobility is found from physical phenomena such as photoconductivity, thermoelectric power, space-charge-limited current, and piezoelectricity [159-161]. These measurement techniques are not generally applicable to a wide range of ceramic materials and, therefore, the reader is referred to the literature for further details.

The Hall effect is one of the more widely used methods for obtaining the mobility of charge-carrying species in ceramics [162, 164]. The Lorentz force equation states that the force $\underline{F}$ on a charge q is given by

$$\underline{F} = q[\underline{E} + \underline{v} \times \underline{B}] \tag{49}$$

where q is the charge and $\underline{v}$ is the velocity of the particle which is subjected to an electric field $\underline{E}$ and a magnetic field $\underline{B}$. If an n-type specimen is subjected to an electric field E_x and magnetic field B_z, as is shown in Fig. 40, a transverse electric field E_y will be set up such that

$$E_y = E_H = \frac{1}{Nq} J_x B_z \tag{50}$$

Here N is the density of charge and J_x is the current density. This is often referred to as the Hall field and

$$V_H = E_y Y \tag{51}$$

the Hall voltage. Here, Y is the dimension of the specimen in the y direction. A commonly defined quantity

$$R_H = \frac{1}{Nq} = \frac{E_y}{J_x B_z} \tag{52}$$

is called the Hall constant, which is negative for electrons and positive for holes. The Hall mobility μ_H is given by

$$\mu_H = R_H \sigma \tag{53}$$

where σ is the electrical conductivity of the material in question.

EXPERIMENTAL TECHNIQUES   155

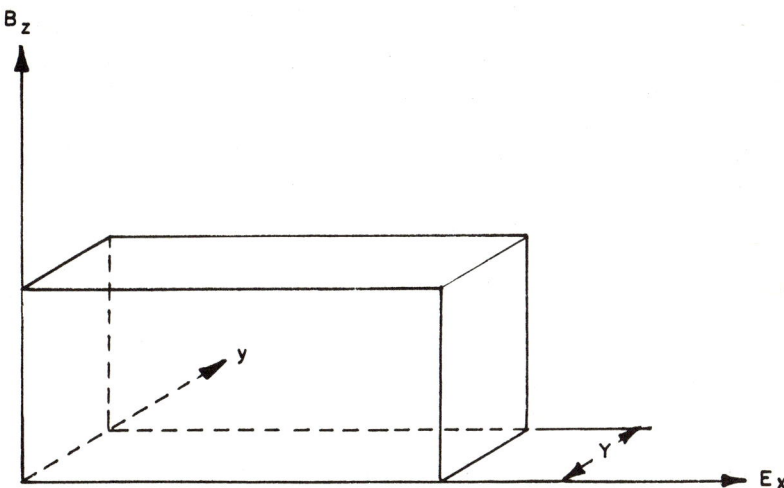

FIG. 40. Orientation of Hall effect specimen in the electric and magnetic fields.

Ceramic materials usually have low electronic mobilities, which would lead to small Hall voltages. Since small voltages are expected, shielding from external noise as described previously will be required. The sensitivity and input impedance of voltage measuring apparatus requires that the Hall voltage be as large as possible and the impedance between the Hall voltage measuring probes be as small as possible. From Eqs. (51) and (52) it is seen that the Hall voltage can be maximized with respect to geometry by having a specimen of small dimension in the x direction and large dimension in the y direction [165]. In reality these dimensional specifications are limited by the impedance of the measuring apparatus, in that the impedance between the Hall probes should be less than that of the measuring apparatus to avoid instrument loading. If the specimen were too thin it would be difficult to handle.

Before considering the measurement of Hall mobilities, several stray effects must be considered [162,166]. It is to be noted that the various measurement techniques to be discussed below are designed to eliminate or minimize the influence of these stray effects.

In placing electrode contacts for the measurement of Hall voltage on a specimen it is difficult if not virtually impossible to insure that they are placed on the same equipotential contours at $B_Z = 0$. This will lead to a voltage V_i between the Hall voltage measuring probes that is a function of the current passing through the sample. Thermoelectric voltages due to the specimen being in a temperature gradient can also lead to a voltage V_T between the Hall probes. Thermal noise, as previously discussed, might also exist at the Hall-voltage probes. This will be troublesome when working with low-conductivity materials at elevated temperatures.

If a current i_x flows through the Hall specimen heat can be absorbed at one current contact and liberated at the other due to the Peltier effect. This can lead to a thermal gradient $\Delta T/\Delta x$ along the sample. This in turn can lead to a diffusion of electronic carriers down this temperature gradient. These carriers can be acted upon by the magnetic field and contribute to the Hall effect. This is referred to as the Nernst effect and leads to a voltage V_N between the Hall probes.

The carriers composing the current i_x will have a distribution of velocities and energies. Those carriers of higher velocity and energy will experience a larger Lorentz force in the y direction, while those of lower velocity and energy will experience a smaller force. This will lead to an accumulation of more energetic electrons at one Hall terminal while the lower-velocity carriers will tend to remain distributed throughout the sample away from this Hall terminal. The resulting temperature gradient gives rise to a thermoelectric voltage V_E between the Hall voltage measuring probes. This effect is referred to as the Ettinghausen effect.

The current resulting from the Nernst effect can also contain charge carriers of varying velocities and energies, which in turn can be acted upon in a similar fashion as in the Ettinghausen effect and thus lead to an additional thermoelectric voltage. This is referred to as the Righi-Leduc effect and leads to a voltage V_{RL} between the Hall terminals.

Distinct from the aforementioned parasitic voltages, the phenomenon of magnetoresistance can lead to errors in the measured values of Hall mobility. The presence of a magnetic field causes the charge carriers to travel in a curved rather than a straight path. There exists a distribution of charge-carrier velocities, carriers of highest velocity experiencing the highest deflecting Lorentz force and leading to the Hall voltage. Thus the x component of velocity will be reduced, with the net effect that the conductivity of the material decreases. This effect is most pronounced on materials with high mobilities subjected to very large magnetic fields. Most ceramic materials have large carrier effective masses and relatively small charge-carrier mobilities, so magnetoresistance should not strongly influence the measured values of mobility. If it is found that the measured values of mobility are effected by the magnitude of the magnetic field, then the conductivity used in Eq. (53) to calculate the Hall mobility might have to be corrected for the magnetoresistive effect.

The simplest, although not necessarily most satisfactory method of measuring the Hall mobility would be via dc techniques. Several variations are used in practice, the main ideas of which will be explored here [166-169].

A lead configuration of Fig. 41 can be used on the specimen to be investigated. Probes I_1 and I_2 are used to supply current to the specimen. Probe pairs V_1 and V_2 are used to measure the voltage required for a four-probe conductivity calculation, while probes V_2 and V_3 are used to measure the Hall voltage [see Eq. (51)]. The specimen is loaded into a constant-temperature region (i.e., a furnace for measurements above room temperature or in a cryostat for lower-temperature measurement) between the pole faces of a magnet. Good measurement practice requires that the specimen and circuitry be enclosed in a grounded shield. Four-probe conductivity measurement at H = 0 are performed as previously described. The magnetic field is turned on and a series of four measurements is performed for the four combinations of current and field polarity, $I(\pm)H(\pm)$, yielding:

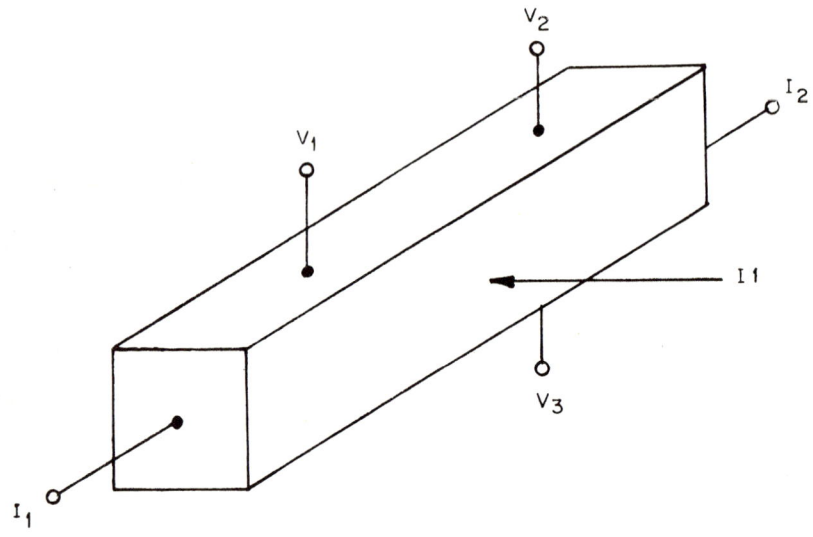

FIG. 41. Lead configuration for Hall-effect specimen.

$$V_a = V_H + V_E + V_N + V_{RL} + V_I \qquad \text{for} \quad I(+), H(+)$$

$$V_b = -V_H - V_E + V_N + V_{RL} - V_I \qquad \text{for} \quad I(-), H(+)$$

$$V_c = V_H + V_E - V_N - V_{RL} - V_I \qquad \text{for} \quad I(-), H(-)$$

$$V_d = V_H - V_E - V_N - V_{RL} + V_I \qquad \text{for} \quad I(+), H(-)$$

The current and magnetic field are reversed and measurements made in a time less than the thermal relaxation time of the specimen. It follows that

$$V_H + V_E = \frac{V_a - V_b + V_c - V_d}{4} \qquad (54)$$

The quantity V_E has not been eliminated by this procedure but is usually smaller than V_H. Theoretically, V_E can be eliminated if the specimen leads are of the same material as the specimen or if ac measurements are performed (as will be dealt with later).

A minor variation of this technique allows much of the mismatch voltage between the Hall probes at $H = 0$ to be eliminated. This variation is depicted in two forms in Fig. 42.

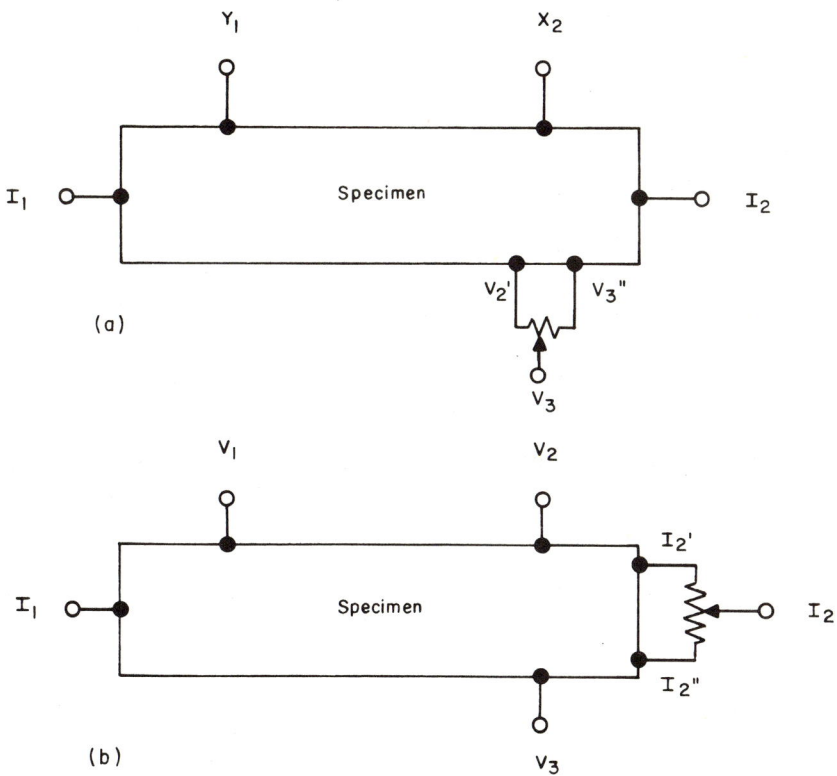

FIG. 42. Alternate lead configurations that can be used to nullify the mismatch voltage at the Hall probes.

In Fig. 42(a) the Hall probe V_3 of Fig. 41 has been replaced by two probes V_3' and V_3'', which are connected to a tapped resistor. The total resistance of this resistor is much greater than that of the specimen between probes V_3' and V_3''. The tap position is adjusted such that the voltage between a and b is zero at $H = 0$. The configuration shown in Fig. 42(b) is better adapted to low-conductivity samples, since the tapped resistor may have a smaller total resistance.

The above method and the aforementioned van der Pauw technique are, due to their simplicity, the principal dc methods used to measure Hall mobility in solids. If proper grounding, guarding, and shielding are used, satisfactory results can be obtained in materials having mobilities greater than about 1.0 cm^2/V-sec. It is noted that magnetoresistance, thermal noise, and the Ettinghausen effect could lead to errors if they were not considered.

Hall mobility techniques using ac electric and magnetic fields can aid in measuring the mobility on materials having low mobilities. These ac techniques are either single ac methods [170,171], where only the electric field is varied, or double ac methods, where both the electric and magnetic fields are varied. In a single ac method the electric rather than the magnetic field is usually varied. The basic type of circuit used for this method is shown in Fig. 43. Here an ac current is supplied to the specimen via an appropriate power supply, which also powers a bucking unit. The bucking unit is basically a phase-shift attenuator. The experimental procedure requires that the conductivity of the specimen be measured by one of the previously described techniques with H = 0. Also with H = 0 the mismatch voltage between the Hall probes V_{H_1} and V_{H_2} is bucked with the aid of the bucking unit. The magnetic field is turned on and the Hall voltage is then measured. It is noted that the tuned detector and transformer can be replaced by a phase-sensitive detector with a differential input if further noise discrimination is required. Lavine has used a single ac method to measure the Hall coefficient in magnetic materials by adding a second bucking unit [171].

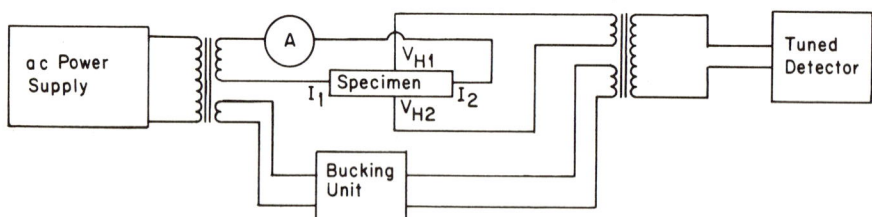

FIG. 43. Circuit for the single ac Hall method.

Two distinct double ac methods are used when measuring the Hall coefficient of materials having low mobilities and/or high resistiv-

EXPERIMENTAL TECHNIQUES   161

ities [172-175]. Both of these are intended to minimize the influence of thermal noise.

The essential points of the double ac method as developed by Hermann and Ham [172] are shown in Fig. 44. An ac magnetic field at a frequency ω_B is produced by rotating the specimen in a dc magnetic field. An ac electric current at a frequency ω_i is slightly less than ω_B. These two frequency components are kept in constant phase with the line frequency, since phase-sensitive detection is used. The Hall signal, being proportional to the product of the magnetic field and current, appears at the frequencies $\omega_B \pm \omega_i$. The frequency of detection used is the higher frequency in order to minimize noise. The frequencies were chosen such that interference from the line frequency would be minimized. Again bucking voltages were used to minimize the misalignment voltage at the Hall probes. With this method and by the use of a field-effect transistor preamplifier, Hall mobilities as small as 0.2 cm^2/V-sec in samples of resistance up to 10^9 ohms could be measured.

A second double ac method has been developed by Lupu, Tallan, and Tannhauser and is represented in Fig. 45. An alternating current at 510 Hz was impressed on the specimen via the internal oscillator of phase-sensitive detector I and the power amplifiers. An ac magnetic field at approximately 2 Hz was established by driving the magnet with a low-frequency power supply. It is observed that in this method, as opposed to the previous one, the frequency of the magnetic field varies significantly from that of the electric field. Therefore, instead of a single phase-sensitive detector being locked on a frequency $\omega_i \pm \omega_B$, two series detectors are used to analyze the resultant wave form (see Fig. 46).

The measurement procedure requires that the conductivity of the specimen be measured with H = 0, so the mismatch voltage across the Hall probes is balanced using the bucking unit. The alternating magnetic field is then turned on. A Hall voltage is developed, as depicted in Fig. 46(a) and is fed into the first phase-sensitive

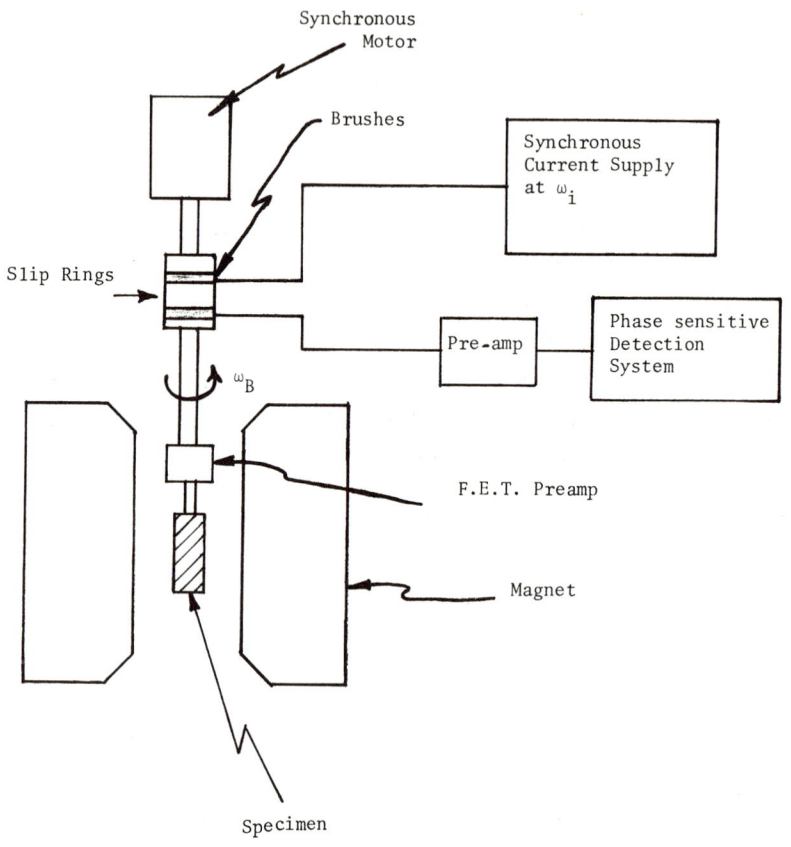

FIG. 44. Experimental system used for the double ac method of Hermann and Ham [172].

detector. The output of this detector, which is shown in Fig. 46(b) is fed to a second phase-sensitive detector which is locked to the frequency and phase of the varying magnetic field via a Halltron probe. The constant value read from this detector is proportional to the Hall voltage developed across the specimen.

The double ac method is effective in minimizing the influence of the thermal noise that is present when measurements are made on materials having low mobilities (as low as 3×10^{-4} cm^2/V-sec) at elevated temperatures (up to 1200°C). Numerous difficulties are

EXPERIMENTAL TECHNIQUES   163

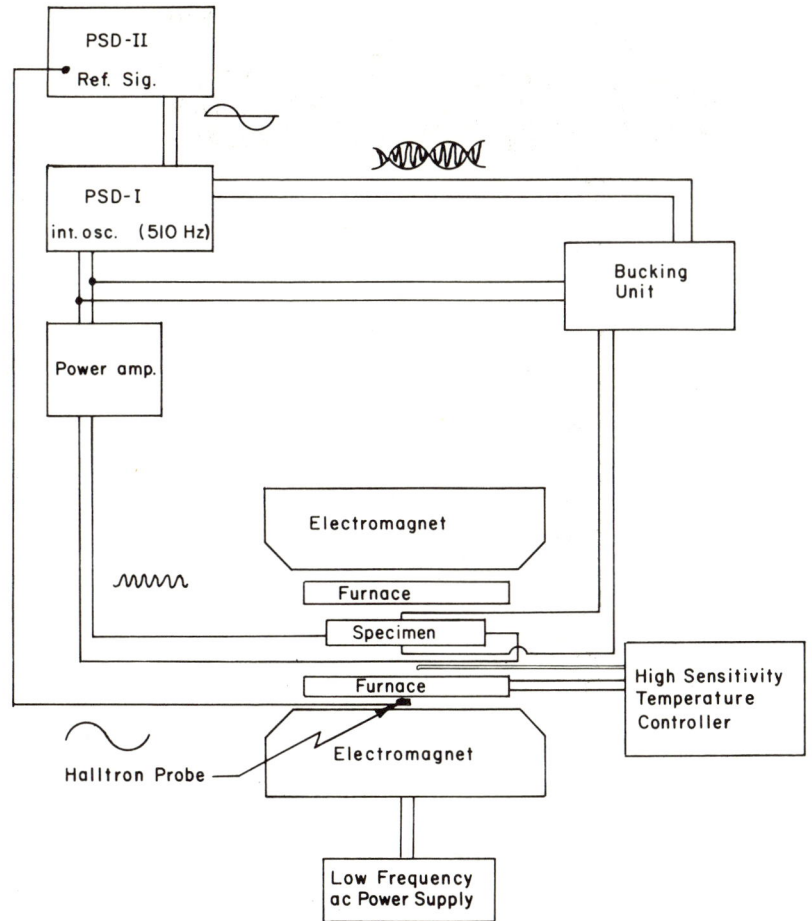

FIG. 45. Experimental system used for the double ac method of Lupu, Tallan, and Tannhauser [174].

encountered when using this technique; these are discussed by McKinzie and Tannhauser [175].

An adaptation of the two aforementioned double ac methods is being used by the present authors [176]. This is schematically represented in Fig. 47. Here the varying magnetic field is obtained by rotating the specimen in a static magnetic field. This has two advantages. It allows a wider range of frequencies of the magnetic

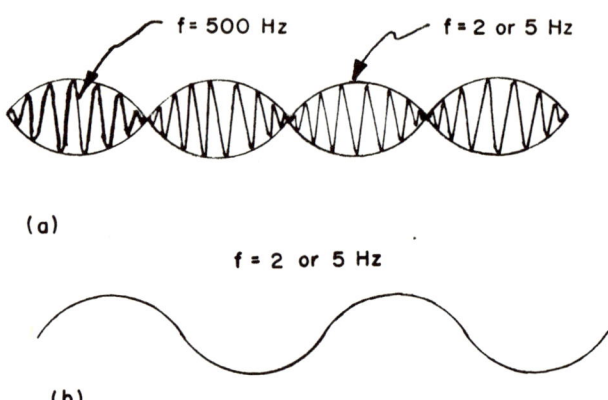

FIG. 46. Wave form as it enters PSD-I (a) and PSD-II (b).

field to be chosen and it allows large pole-face spacing of the magnet and a larger specimen chamber and furnace to be used. The bucking signal used to balance the mismatch voltage between the Hall probes is obtained from the probe pairs V_1 and V_2. Thus any temperature changes or other variations in the system that might change the mismatch voltage level are automatically compensated.

In addition to the experimental difficulties encountered when measuring the Hall mobility of charge carriers in materials, the nature of the carriers and their associated scattering mechanisms must be considered when analyzing the experimental results [163]. The Hall coefficient is related to the carrier concentrations n and p via the relationship

$$R = \frac{r}{e} \frac{p\mu_p^2 - n\mu_n^2}{(p\mu_p + n\mu_n)^2} \tag{55}$$

Here e is the electron charge and r is a numerical factor which varies between about 1 and 2 according to the type of scattering which predominates and also with the degree of degeneracy in the conduction or valence band. The quantities μ_p and μ_n are the hole and electron mobilities, respectively. If the material is n-type, Eq. (55) reduces to

$$R = \frac{-r}{ne} \tag{56}$$

EXPERIMENTAL TECHNIQUES

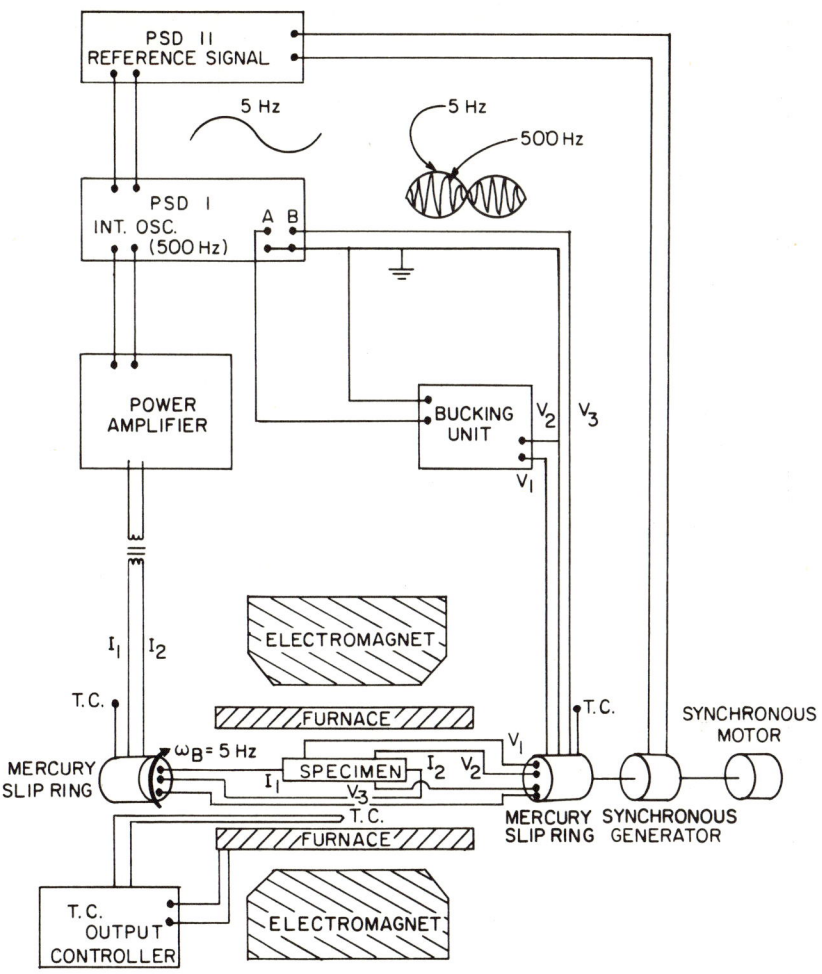

FIG. 47. Experimental system used for the double ac method of Seitz [176].

whereas for a p-type material

$$R = \frac{+r}{pe} \tag{57}$$

For other, nonlimiting cases Eq. (55) would have to be used to interpret the results obtained.

Although the Hall methods are most often used in measuring the charge-carrier mobilities in ceramic materials, other more limited techniques are sometimes useful. The conductivity mobility can be determined from the expression

$$\sigma = ne\mu_c \tag{58}$$

if the carrier concentration n is known. Two means of determining the carrier concentration are available. At low to moderate temperatures and over a limited temperature range a saturation region may exist if an appropriate impurity doping is introduced into the material being investigated [163]. For this method to be used a known homogeneous concentration of the selected impurity would have to be incorporated in the specimen. A second means of determining the concentration of charge carriers is through the use of thermogravimetric data [177]. In order to use this method the concentration of point defects, which would have to have a single state of ionization, is inferred from weight-change data. Also, this method can only be used over ranges of temperatures where the state of ionization of defects is known.

The drift mobility of charge carriers can be measured on ceramic materials if the appropriate conditions exist. Two basic methods are available to measure the drift mobility of charge carriers in solids. The first is shown in Fig. 48 [159,162,178]. Here, the specimen under investigation is subject to an applied electric field. A dc supply may be used if space-charge buildup is unimportant, otherwise a pulsed voltage source must be used. An excess concentration of the minority charge carriers are injected at contact e using a pulsed voltage source. Photoexcitation can also be used to inject free carriers if the specimen is photosensitive. These excess carriers drift down the specimen under the influence of an electric field $\underline{E}$ and are collected at electrode c, which is a distance L from electrode e. The pattern shown on the oscilloscope is developed due to the injection and motion of these minority carriers. The first pulse is the result of electromagnetic interaction upon carrier injection, while the second pulse occurs when the charge carriers

EXPERIMENTAL TECHNIQUES   167

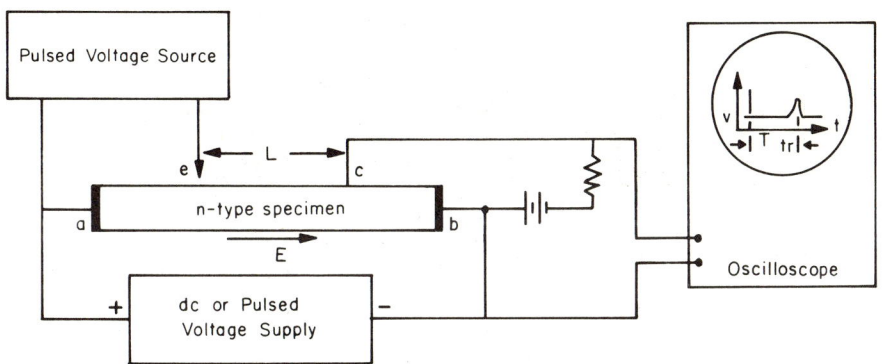

FIG. 48. Circuit for the measurement of drift mobility in materials having large charge-carrier mobilities.

reach electrode c. The carrier mobility, which is the velocity per unit field, is given by

$$\mu_d = \frac{L}{E\,T_{tr}} \tag{59}$$

where T_{tr} is the time required for the injected charge carriers to travel from the injection probe to the collection probe. Several requirements must be fulfilled if this method is to be applicable to ceramic materials. The two end contacts a and b must be ohmic, while contact e must be injecting. Charge-carrier trapping must be insignificant. If trapping were significant the charge carriers would be trapped before reaching electrode c and the second pulse would not be observed on the oscilloscope. If a minor degree of charge-carrier trapping occurs appropriate corrections must be made [159]. In practice it is found that this technique is applicable to materials having charge-carrier mobilities greater than 500 cm^2/V-sec.

A second means of measuring the drift mobility of charge carriers in ceramics is shown in Fig. 49 [179-183]. Here a pulsed source of ionizing radiation, either electrons or photons, excites electron-hole pairs on one surface of a thin plane parallel and high-resistivity sample. Either the electrons or the holes, depending on the

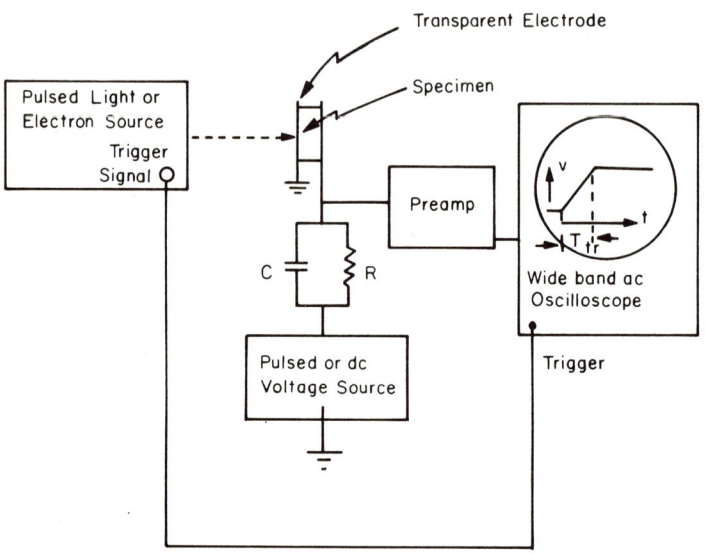

FIG. 49. Circuit for the measurement of drift mobilities in low-mobility materials.

polarity of the applied bias, are drawn across the sample under the influence of the applied electric field. As these carriers move across the sample they induce charge on the back surface of the sample. This charge is integrated via the RC network and recorded on the oscilloscope. The pattern shown on the oscilloscope is as would occur for the transit of holes. The charge-carrier mobility may be obtained from Eq. (59) by using the measured transit time T_{tr}, applied field E, and specimen thickness L. This method is well adapted to the measurement of drift mobilities in the range 10^1 to 10^{-5} cm^2/V-sec in photosensitive materials which have high resistivities. If charge-carrier trapping were important, appropriate corrections would, again, be required [179-183].

REFERENCES

1. J. W. Cleland, "Purity and Perfection of Research Specimens of Oxides," in Mass Transport in Oxides (J. B. Wachtman and A. D. Franklin, eds.), NBS Special Pub. 296, U.S. Govt. Printing Office, Washington, D.C., 1968, p. 195.

2. G. S. Brady, *Materials Handbook*, McGraw-Hill, New York, 1971.

3. "The Characterization of Materials," Materials Advisory Board, Report MAB 299, 1967.

4. *Engineering Properties of Selected Ceramic Materials*, compiled and edited by Battelle Memorial Institute, American Ceramic Society, 1966.

5. F. A. Kroger, *Chemistry of Imperfect Crystals*, North-Holland, Amsterdam, 1964.

6. H. S. Peiser, ed., *Crystal Growth*, Pergamon Press, New York, 1966.

7. H. Schafer, *Chemical Transport Reactions*, Academic Press, New York, 1964.

8. N. N. Sheftal, ed., *Growth of Crystals*, Consultants Bureau, New York, 1969.

9. P. F. Kane and G. B. Larrabee, *Characterization of Semiconductor Materials*, McGraw-Hill, New York, 1970.

10. G. G. Koerber, *Properties of Solids*, Prentice-Hall, Englewood Cliffs, New Jersey, 1962.

11. B. D. Cullity, *Elements of X-Ray Diffraction*, Addison-Wesley, Reading, Mass., 1959.

12. W. D. Kingery, *Introduction to Ceramics*, Wiley, New York, 1960.

13. W. D. Kingery, ed., *Kinetics of High-Temperature Processes*, Part IV, MIT Press, Cambridge, Mass., John Wiley and Sons, New York, and Chapman and Hall, London, 1959.

14. A. U. Seybolt and J. E. Burke, *Procedures in Experimental Metallurgy*, John Wiley and Sons, New York, 1953.

15. R. M. Spriggs, "Hot-Pressed Oxides," in *High Temperature Oxides*, Part III (A. M. Alper, ed.), Academic Press, New York, 1970.

16. T. H. Etsell and S. N. Flengas, *Chem. Rev.*, $\underline{70}$, 339 (1970).

17. D. Miliotis and D. N. Yoon, *J. Phys. Chem. Solids*, $\underline{30}$, 1241 (1969).

18. J. Volger, *Progress in Semiconductors*, Vol. 4 (A. F. Gibson, ed.), Wiley, New York, 1960, p. 206.

19. D. F. Gibbs and B. W. Jones, *Brit. J. Appl. Phys. D*, $\underline{3}$, 157 (1970).

20. R. H. Bube, Photoconductivity of Solids, John Wiley and Sons, New York, 1960.

21. M. J. O. Strutt, Semiconductor Devices, Academic Press, New York, 1966, p. 110-118.

22. J. Lindmayer and C. Y. Wrigley, Semiconductor Devices, D. Van Nostrand, Princeton, New Jersey, 1965, p. 30.

23. K. Kiukkola and C. Wagner, J. Electrochem. Soc., 104, 308; 379 (1957).

24. B. G. H. Steele, in Electromotive Force Measurements in High-Temperature Systems (C. B. Alcock, ed.), American Elsevier, New York, 1968, p. 3.

25. T. L. Markin, in Electromotive Force Measurements in High-Temperature Systems (C. B. Alcock, ed.), American Elsevier, New York, 1968, p. 91.

26. B. C. H. Steele, in Mass Transport in Oxides (J. B. Wachtman, Jr. and A. D. Franklin, eds.), NBS Special Pub. 296, 1968, p. 165.

27. B. C. H. Steele and C. B. Alcock, Trans. Met. Soc. AIME, 233, 1359 (1965).

28. D. O. Raleigh, in Progress in Solid State Chemistry, Vol. 3, Pergamon, Oxford, 1966, p. 83.

29. R. A. Rapp and D. A. Shores, Techniques in Metal Research, Vol. 4, Chap. VI C, Wiley-Interscience, New York, 1970.

30. G. B. Barbi, J. Phys. Chem., 68, 1025 (1964).

31. F. E. Rizzo, L. R. Bidwell, and D. F. Frank, Trans. Met. Soc. AIME, 239, 1901 (1967).

32. C. B. Alcock and S. Zador, Electrochim. Acta., 12, 673 (1967).

33. R. N. Blumenthal and D. H. Whitmore, J. Amer. Ceram. Soc., 44, 508 (1961).

34. R. N. Blumenthal and D. H. Whitmore, J. Electrochem. Soc., 110, 92 (1963).

35. R. N. Blumenthal, J. B. Moser, and D. H. Whitmore, J. Amer. Ceram. Soc., 48, 617 (1965).

36. Susan Zador, in Electromotive Force Measurements in High-Temperature Systems (C. B. Alcock, ed.), American Elsevier, New York, 1968, p. 145.

37. K. Kiukkola, Acta Chem. Scand., **16**, 327 (1962).

38. F. A. Kuznetsov, V. I. Belii, and T. N. Rezukhina, Dokl. Akad. Nauk SSSR Ser. Fiz-Khim., **139**, 1405 (1961).

39. H. F. Rizzo, R. S. Gordon, and I. B. Cutler, in Mass Transport in Oxides (J. B. Wachtman, Jr., and A. D. Franklin, eds.), NBS Special Pub. 296, U.S. Govt. Printing Office, Washington, D.C., 1968, p. 129.

40. Yuri D. Tretyakov and Robert A. Rapp, Trans. Met. Soc. AIME, **245**, 1235 (1969).

41. T. H. Etsell and S. N. Flengas, Chem. Rev., **70**, 341 (1970).

42. J. E. Bauerle, J. Chem. Phys., **45**, 4162 (1966).

43. John B. Hardaway III, John W. Patterson, D. R. Wilder, and Jon D. Schieltz, J. Amer. Ceram. Soc., **54**, 94 (1971).

44. C. B. Alcock, S. Zador, and B. C. H. Steele, Proc. Brit. Ceram. Soc., **8**, 231 (1967).

45. H. Schmalzried, Z. Phys. Chem. (Frankfurt), **29**, 77 (1961).

46. T. C. Markin and R. J. Bones, U. K. Atomic Energy Authority, Report No. 4178.

47. C. B. Alcock, S. Zador, and B. C. H. Steele, Proc. Brit. Ceram. Soc., **8**, 231 (1967).

48. O. Kubaschewski, E. Ll. Evans, and C. B. Alcock, Metallurgical Thermochemistry, 4th Ed., Pergamon Press, 1967, p. 157.

49. W. C. Tripp, R. W. Vest, and N. M. Tallan, in Vacuum Microbalance Techniques, Vol. 4 (P. M. Walters, ed.), Plenum Press, New York, 1965, p. 141.

50. W. Kreisman, Ultra Microweight Determination in Controlled Environments (S. P. Wolsky and E. J. Zdanuk, eds.), Wiley-Interscience, 1969, p. 163.

51. J. Weissbart and R. Ruka, Rev. Sci. Instr., **32**, 593 (1961).

52. L. S. Karken and R. W. Gurry, Physical Chemistry of Metals, McGraw-Hill, New York, 1953, p. 352.

53. C. E. Wicks and F. E. Block, Bureau of Mines, Bulletin No. 605, U.S. Govt. Printing Office, Washington, D.C., 1963.

54. T. L. Markin, R. J. Bones, and V. J. Wheeler, Proc. Brit. Ceram. Soc., **8**, 51 (1967).

55. D. J. M. Bevan and J. Kordis, J. Inorg. Nucl. Chem., 26, 1509 (1964).

56. R. W. Vest, "The Electrical Conductivity of Ceramics. Its Measurement and Interpretation," Office of Aerospace Research, ARL Report 67-0010.

57. L. S. Darken and R. W. Gurry, J. Amer. Ceram. Soc., 67, 1398 (1945).

58. G. G. Libowitz, Energetics in Metallurgical Phenomena, Vol. 4 (W. M. Mueller, ed.), Gordon and Breach, New York, 1968, p. 73.

59. A. B. Scott and M. E. Hills, J. Chem. Phys., 28, 24 (1958).

60. G. G. Libowitz, J. Amer. Chem. Soc., 15, 1501 (1953).

61. B. Fisher and D. S. Tannhauser, J. Electrochem. Soc., 111, 1194 (1964).

62. N. G. Eror, Jr., Ph.D. Thesis, Northwestern University, 1965.

63. P. Kofstad, Phys. Chem. Solids, 23, 1579 (1962).

64. K. Forland, Acta Chem. Scand., 18, 1267 (1964).

65. P. Kofstad and P. B. Anderson, Phys. Chem. Solids, 21, 280 (1961).

66. W. C. Tripp and N. M. Tallan, J. Amer. Ceram. Soc., 53, 531 (1970).

67. R. J. Panlener and R. N. Blumenthal, J. Amer. Ceram. Soc., 54, 610 (1971).

68. D. C. Fox and M. J. Katz, in Ultra Microweight Determination in Controlled Environments (S. D. Wolsky and E. J. Zdanuk, eds.) Wiley-Interscience, 1969, p. 465.

69. M. J. Katz, ed., Vacuum Microbalance Techniques, Vol. 1, Plenum Press, New York, 1961.

70. R. F. Walker, ed., Vacuum Microbalance Techniques, Vol. 2, Plenum Press, New York, 1962.

71. K. H. Berndt, ed., Vacuum Microbalance Techniques, Vol. 3, Plenum Press, New York, 1963.

72. P. M. Waters, ed., Vacuum Microbalance Techniques, Vol. 4, Plenum Press, New York, 1964.

73. K. H. Berndt, ed., Vacuum Microbalance Techniques, Vol. 5, Plenum Press, New York, 1965.

74. A. W. Czanderna, ed., Vacuum Microbalance Techniques, Vol. 6, Plenum Press, New York, 1966.

75. C. H. Massen and H. J. van Beckum, eds., Vacuum Microbalance Techniques, Vol. 7, Plenum Press, New York, 1970.

76. A. W. Czanderna, ed., Vacuum Microbalance Techniques, Vol. 8, Plenum Press, New York, 1971.

77. A. W. Czanderna, in Ultra Microweight Determination in Controlled Environments (S. P. Wolsky and E. J. Zdanuk, eds.), Wiley-Interscience, 1969, p. 7.

78. W. C. Tripp, R. W. Vest, and H. C. Graham, in Vacuum Microbalance Techniques, Vol. 6 (A. W. Czanderna, ed.) Plenum Press, New York, 1967, p. 107.

79. G. Greskovich and H. Schmalzried, J. Phys. Chem. Solids, 31, 639 (1970).

80. R. H. Bube, Photoconductivity of Solids, John Wiley and Sons, New York, 1960.

81. A. Rose, Proc. IRE, 43, 1850 (1955).

82. R. L. Petritz, in Photoconductivity, Wiley, New York, 1956, p. 49.

83. F. L. Lummis and R. L. Petritz, Phys. Rev., 105, 502 (1957).

84. K. M. Van Vliet and J. Blok, Physica, 22, 525 (1956).

85. K. M. Van Vliet, J. Blok, X. Ris, and X. Steketee, Physica, 22, 723 (1956).

86. X. Pearson, X. Montgomery, and X. Feldmann, J. Appl. Phys., 27, 91 (1956).

87. T. S. Moss, Proc. IRE, 43 1869 (1955).

88. C. I. Shulman, Phys. Rev., 98, 384 (1955).

89. W. Shockley, Electrons and Holes in Semiconductors, Van Nostrand, Princeton, New Jersey, 1950.

90. Keithley Instruments Catalog, Keithley Instruments, Inc., Cleveland, Ohio, 1970.

91. *Elimination of Noise in Low-Level Circuits*, Clevite Corp.

92. E. Frank, *Electrical Measurement Analysis*, McGraw-Hill, New York, 1959.

93. M. A. Seitz and D. H. Whitmore, *J. Phys. Chem. Solids*, 29, 1033 (1968).

94. M. A. Seitz, R. T. McSweeney, and W. M. Hirthe, *Rev. Sci. Instrum.*, 40, 826 (1969).

95. *DC Resistance or Conductance of Insulating Materials*, American Society for Testing and Materials, Philadelphia, Pa., 1967.

96. A. J. Moulson and P. Popper, *Proc. Brit. Ceram. Soc.*, 10, 41 (1968).

97. R. W. Wallace and E. Ruh, *J. Amer. Ceram. Soc.*, 50, 358 (1967).

98. R. R. Schemmel and R. L. Gordon, Paper presented at 73rd Annual Meeting of the American Ceramics Society, Chicago, Ill., Apr. 26-28, 1971.

99. R. R. Schemmel, R. L. Gordon, and J. L. Bates, *ANS Trans.*, 13, 800 (1970).

100. S. P. Mitoff, *J. Chem. Phys.*, 41, 2561 (1964).

101. T. M. Dauphinee and E. Mooser, *Rev. Sci. Instrum.*, 26, 660 (1955).

102. L. B. Valdes, *Proc. IRE*, 42, 420 (1954).

103. F. M. Smits, *Bell Systems Tech. J.*, May (1958).

104. L. J. Van der Pauw, *Philips Res. Rept.*, 13, 1 (1958).

105. R. N. Blumenthal, W. M. Hirthe, and B. A. Pinz, in *Anisotropy in Single-Crystal Refractory Compounds*, Vol. 1 (F. W. Vahldiek and S. A. Mersol, eds.), Plenum Press, New York, 1968.

106. C. Wagner, *Z. Phys. Chem. (Leipzig)*, B21, 42 (1933); B23, 469 (1934).

107. M. H. Hebb, *J. Chem. Phys.*, 20, 185 (1952).

108. F. A. Kröger, *Chemistry of Imperfect Crystals*, Chap. 22, North-Holland, Amsterdam, 1964.

109. C. Wagner, in Thermodynamics and Kinetics, Proc. 7th Meeting Int. Comm. on Electrochem., Lindau (1955), Butterworths, London, 1957, p. 361.

110. X. Nagel and C. Wagner, Z. Phys. Chem. (Leipzig), B25, 71 (1934).

111. C. Tubandt, in Handbuch der Experimentalphysik (X. Wies and X. Harm, eds.), Vol. 12, Part II, Akademische Verlagsgesellschaft, Leipzig, 1933, p. 412.

112. A. B. Lidiard, in Handbuch der Physik, Vol. 20, Springer-Verlag, Berlin, 1957.

113. F. Kerkhoff, Z. Phys., 130, 449 (1951).

114. R. J. Brook, W. L. Pelzmann, and F. A. Kröger, J. Electrochem. Soc., 118, 185 (1971).

115. H. Yanagida, R. J. Brook, and F. A. Kröger, J. Electrochem. Soc., 117, 593 (1970).

116. L. Heyne, "Ionic Conductivity in Oxides," in Mass Transport in Oxides (J. B. Wachtman and A. D. Franklin, eds.), NBS Special Pub. 296, U.S. Govt. Printing Office, Washington, D.C., 1968, pp. 149-164.

117. C. Wagner, Z. Phys. Chem., B21, 25 (1933).

118. C. B. Alcock and G. P. Stavropoulos, J. Amer. Ceram. Soc., 54, 436 (1971).

119. H. Schmalzried, J. Chem. Phys., 33, 940 (1960).

120. C. F. Pal'guev and A. D. Neuimis, Soviet Phys. — Solid State, 4, 629 (1962).

121. S. P. Mitoff, J. Chem. Phys., 36, 1383 (1962).

122. S. P. Mitoff, J. Chem. Phys., 41, 2561 (1964).

123. H. Schmalzried, Z. Phys. Chem. (Frankfurt), 38, 87 (1963).

124. J. B. Hardaway III, J. W. Patterson, D. R. Wilder, and J. D. Schieltz, J. Amer. Ceram. Soc., 54, 94 (1971).

125. R. J. Brook, J. Yee, and F. A. Kröger, J. Amer. Ceram. Soc., 54, 444 (1971).

126. G. Van Handel, Ph.D. Thesis, Marquette University, 1972.

127. R. N. Blumenthal and D. H. Whitmore, J. Electrochem. Soc., 110, 92 (1963).

128. D. W. Peters, L. F. Feinstein, and Christian Peltzer, J. Chem. Phys., 42, 2345 (1964).

129. J. P. Loup and A. M. Anthony, Rev. Hautes Temp. Refract., 1, 15 (1964).

130. A. J. Moulson and P. Popper, Proc. Brit. Ceram. Soc., 10, 41 (1968).

131. O. T. Ozkan and A. J. Moulson, Brit. J. Appl. Phys., 3, 983 (1970).

132. B. Ilschner, J. Chem. Phys., 28, 1109 (1958).

133. J. B. Wagner and C. Wagner, J. Electrochem. Soc., 104, 509 (1957).

134. J. B. Wagner and C. Wagner, J. Chem. Phys., 26, 1597 (1957).

135. John W. Patterson, E. C. Bogren, and Robert A. Rapp, J. Electrochem. Soc., 114, 752 (1967).

136. C. Wagner, Z. Elektrochem., 60, 4 (1956); 63, 1027 (1959).

137. D. O. Raleigh, in Progress in Solid State Chemistry, Chap. 3, Pergamon Press, Oxford, 1967.

138. D. O. Raleigh, J. Phys. Chem. Solids, 26, 329 (1965).

139. R. Vest and N. Tallan, J. Appl. Phys., 36, 543 (1965).

140. J. M. Wimmer, L. R. Bidwell, and N. M. Tallan, J. Amer. Ceram. Soc., 50, 198 (1967).

141. W. Danforth and J. Bodine, J. Franklin Inst., 260, 467 (1955).

142. R. E. Howard and A. B. Lidiard, Discussions Faraday Soc., 23, 113 (1957).

143. R. W. Howard and A. B. Lidiard, Rept. Progr. Phys., 27, 161 (1964).

144. R. R. Heikes and R. W. Ure, Thermoelectricity, Wiley-Interscience, New York, 1961.

145. N. Cusack and P. Kendall, Proc. Phys. Soc. (London), 72, 898 (1958).

146. F. A. Kroger, The Chemistry of Imperfect Crystals, John Wiley and Sons, Inc., New York, 1964, pp. 928-940.

147. R. J. Ruka, J. W. Bauerle, and L. Dykstra, J. Electrochem. Soc., 115, 497 (1968).

148. N. M. Tallan and I. Bransky, J. Electrochem. Soc., 118, 345 (1971).

149. J. L. Weininger, J. Electrochem. Soc., 111 (N5), 769 (1964).

150. J. Thiele, Phys. Z., 26, 321 (1925).

151. T. I. Nikitinskaya and A. N. Murin, Zh. Tekhn. Fiz., 25, 1193 (1955).

152. P. W. M. Jacobs and J. N. Maycock, Trans. AIME, 236, 165 (1966).

153. W. R. Thurber and A. J. H. Mante, Phys. Rev., 139, A1655 (1955).

154. A. P. Young, P. B. Robbins, W. B. Wilson, and C. M. Swartz, Rev. Sci. Instrum., 31, 70 (1960).

155. A. R. Allnatt and P. W. M. Jacobs, Proc. Roy. Soc., Ser. A, 267, 31 (1962).

156. H. R. Meddins and J. E. Parrott, Brit. J. Appl. Phys., 2, 691 (1969).

157. N. M. Tallan, private communication.

158. R. J. Brook, J. Lee, and F. A. Kroger, J. Electrochem. Soc., 54, 444 (1971).

159. R. H. Bube, Photoconductivity of Solids, Wiley, New York, 1960.

160. A. Many, S. Z. Weisz, and M. Simhony, Phys. Rev., 126, 1989 (1962).

161. N. I. Meyer, M. H. Jorgensen, and E. Mosekilde, J. Phys. Soc. Jap., 21, 406 (1966).

162. N. Cusek, The Electrical and Magnetic Properties of Solids, Longmans, Green, London, 1958.

163. R. A. Smith, Semiconductors, Cambridge Univ. Press, London, 1959.

164. C. Kittel, Introduction to Solid State Physics, Wiley, New York, 1966.

165. I. Isenberg, B. R. Russell, and R. F. Greene, Rev. Sci. Instrum., 19, 685 (1948).

166. O. Lindberg, Proc. IRE, 40, 1414 (1950).

167. B. R. Russell and C. Wahlig, New Instrum., 1028 (Sept. 1950).

168. T. M. Dauphinee and E. Mooser, Rev. Sci. Instrum., 26, 660 (1955).

169. L. J. van der Pauw, Philips Res. Rept., 13, 1 (1958).

170. S. Fujime, M. Murakami, and E. Hirahara, J. Phys. Soc. Jap., 16, 183 (1961).

171. J. M. Lavine, Rev. Sci. Instrum., 29, 970 (1958).

172. A. M. Hermann and J. S. Ham, Rev. Sci. Instrum., 36 (1965).

173. J. L. Levy, Phys. Rev., 92, 215 (1953).

174. N. Z. Lupu, N. M. Tallan, and D. S. Tannhauser, Rev. Sci. Instrum., 38, 1658 (1967).

175. H. L. McKinzie and D. S. Tannhauser, J. Appl. Phys., 40, 4954 (1969).

176. M. A. Seitz and W. M. Hirthe, Semi-Annual Progress Report, Aerospace Research Lab., Wright-Patterson Air Force Base, ARZ Contract No. F33615-68-C-1355, Oct. 1970 to July 1971.

177. R. N. Blumenthal and R. J. Panlener, J. Phys. Chem. Solids, 31, 1190 (1969).

178. J. R. Haynes and W. Shockley, Phys. Rev., 75, 691 (1949).

179. W. E. Spear, Proc. Phys. Soc., B69, 1139 (1956).

180. W. E. Spear, Proc. Phys. Soc., B70, 669 (1957).

181. W. E. Spear, Proc. Phys. Soc., 76, 828 (1960).

182. E. Hartke, J. Appl. Phys., 36, 3850 (1965).

183. M. A. Seitz and D. H. Whitmore, J. Phys. Chem. Solids, 29, 1033 (1969).

Chapter 3

DEFECT STRUCTURE OF CERAMIC MATERIALS

R. J. Brook

Atomic Energy Research Establishment
Materials Development Division
Harwell
Didcot, Berkshire, U. K.

I.	INTRODUCTION	180
II.	TYPES OF DEFECT	181
	A. Point Defects	181
	B. Other Crystalline Defects	184
III.	ORIGIN OF POINT DEFECTS	185
	A. Defects as Equilibrium Species	185
	B. Notation for Defects and Defect Reactions	191
	C. Intrinsic Atomic Disorder	193
	D. Extrinsic Atomic Disorder	197
	E. Radiation Damage and Plastic Deformation	201
IV.	DEFECTS AND ELECTRON ENERGY LEVELS	201
	A. Band Structure and Defect Levels	201
	B. Intrinsic Electronic Disorder	209
	C. Extrinsic Electronic Disorder	210
	D. Occupation of Levels	211
V.	DEFECT EQUILIBRIA IN CRYSTALS	213
	A. Small Defect Concentrations	213
	B. Large Defect Concentrations	233
	C. Extended Defects	240
VI.	MICROSTRUCTURAL EFFECTS	242

VII.	DEFECTS IN GLASSES	248
VIII.	DEFECTS AND ELECTRICAL CONDUCTIVITY	250
	A. General	250
	B. Ionic Conductivity	252
	C. Electronic Conductivity	253
	D. Mixed Conductivity: Conductivity vs Oxygen Pressure Diagrams	254
	E. Interpretation of Conductivity Measurements	257
IX.	CONCLUSION	264
	APPENDIX	265
	REFERENCES	266

I. INTRODUCTION

A necessary preliminary to the study of electrical conduction phenomena in insulating and semiconducting ceramic materials is a discussion of point defects. In metals, where free electrons are present as a result of the nature of the bonding, conductivity is a feature of the perfect crystal; in semiconductors and insulators, however, charge carriers are to be regarded as defects in the perfect crystalline order, and consideration of charge transport leads necessarily to consideration of point defects and their migration.

Electrical conductivity in ceramics can involve both electronic and ionic charge carriers. In a defect-free insulator or semiconductor the electrons may be considered as either bound to the individual atoms that form the solid or else as located in a filled electron band, the choice of model depending on the particular material. In the first case they are by definition unable to move; in the second case, the filled band means that there are no electron states available whose occupation would correspond to electron displacement in response to an applied field (see Section IV and Chapter 4). Similarly, there is, in a perfect crystal, no mechanism

DEFECT STRUCTURE OF CERAMIC MATERIALS 181

for ionic species to diffuse, and hence no current can result from ionic motion.

In considering electrical conduction in ceramics, then, we are concerned with defects and defect migration. The purpose of this chapter is to study this connection more closely by describing the various types of defect, the ways in which they can be created, and the effect they can have in producing electrical conductivity. The relations between electrical conduction and other variables, such as temperature, sample composition, and microstructure, will then be discussed in terms of the types and concentrations of defects and their response to changes in these variables.

A complementary aim is to examine the interconnection between defects and conductivity from the other side, and to show how measurements of electrical conductivity can be used to reveal the defect structure of a particular material. This can give understanding of the other processes where defects can be important, such as diffusion, sintering, creep, tarnishing, and solid-state reactions.

Ceramic materials will be considered throughout in terms of inorganic, nonmetallic compounds; discussion will be restricted to insulators or semiconductors, and the type of bonding involved may be ionic or covalent, except where specified. Those ceramics which display metallic behavior are not specifically treated here, since defects are less significant in the interpretation of their properties; they are described fully in Chapter 6.

II. TYPES OF DEFECT

A. POINT DEFECTS

Electrical conductivity σ is given by the equation

$$\sigma = \Sigma_a (n_a e_a \mu_a) \tag{1}$$

where n_a is the concentration of charge carrier a, e_a is its charge, and μ_a its mobility (drift velocity in unit potential gradient).

The charge carriers a will generally be point defects or points within the material where departure from the perfect crystalline array has occurred.

We may identify for a pure crystal the following types of atomic point defect:

Type 1: vacancies; these are sites where an atom is missing from a normally occupied position.

Type 2: interstitials; these are sites where an atom is found in a normally unoccupied position.

Type 3: misplaced atoms; these are sites where one type of atom is found at a site normally occupied by another.

For an impure crystal, we may in addition to Types 1-3, identify:

Type 4: impurity atoms; these may occupy normally unoccupied positions (in which case they are termed interstitial) or positions normally occupied by one of the host atoms (in which case they are termed substitutional).

In addition to these atomic defects (see Fig. 1), there are electronic defects:

Type 5: the free electron.
Type 6: the free electron hole.

These six are the basic types of defect important in influencing the electrical conductivity of a material. There are further point defects that result from interaction between these basic types (these are termed associates), as well as point defects of a more elaborate structure than the basic types (these are termed complex defects). These are discussed in Section V,B.

As defined above, point defects of Types 1-4 are described in terms of atoms; for instance, the vacancy is treated as a missing atom. If the crystal is strongly ionic in character, these atomic defects, once formed, will ionize by collecting or losing electrical charges. If we consider as an example an oxygen vacancy in an ionic oxide, then on removing the oxygen atom two electrons are left in

DEFECT STRUCTURE OF CERAMIC MATERIALS   183

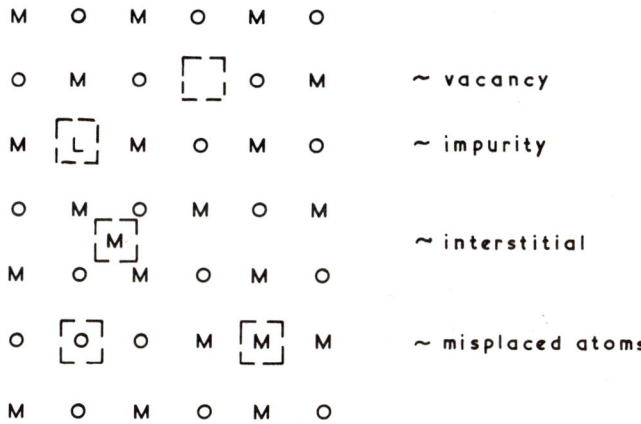

FIG. 1. Schematic picture of atomic point defects in an oxide, MO.

the region of the vacancy [see Fig. 2(a)]. The vacancy can then ionize, with one or both of the electrons leaving the vacancy for lower-energy positions in the lattice. Depending on the particular material, they may become free electrons, or electrons attached to M^{2+} ions [converting them to M^+, as shown in Fig. 2(b)], or may be destroyed by joining electron holes left by the formation of a metal vacancy. The end result of the latter case is the formation of two ion vacancies [see Fig. 2(c)].

Formation of such "ionic" defects may be formally described in two steps: first the creation of an atomic defect, and then the ionization of the defect. In ceramic systems such ionization probably occurs more often than not, but the exact situation will depend, as discussed later (Section IV), on the electron energy-level scheme. In the following, "ionic" defects will be termed "fully ionized atomic defects," a notation which is convenient where defect concepts are applied to different materials with different bonding types.

It should be noted that ionization as discussed here involves movement of electrons from higher to lower energy levels; another situation where electrons can leave an atomic defect occurs when

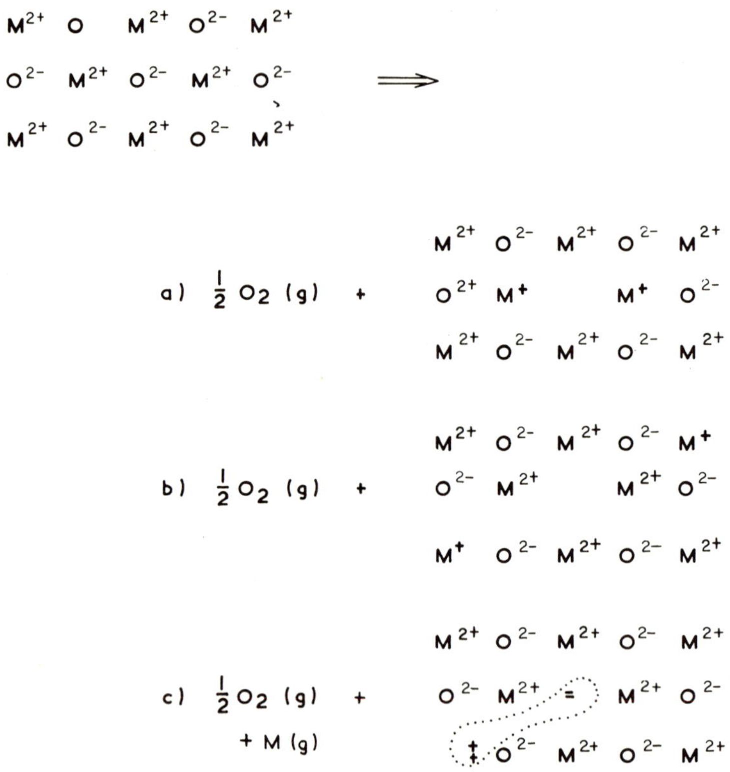

$$M^{2+} \quad O \quad M^{2+} \quad O^{2-} \quad M^{2+}$$
$$O^{2-} \quad M^{2+} \quad O^{2-} \quad M^{2+} \quad O^{2-} \quad \Longrightarrow$$
$$M^{2+} \quad O^{2-} \quad M^{2+} \quad O^{2-} \quad M^{2+}$$

$$M^{2+} \quad O^{2-} \quad M^{2+} \quad O^{2-} \quad M^{2+}$$
a) $\frac{1}{2} O_2 (g) \quad + \quad O^{2+} \quad M^+ \qquad M^+ \quad O^{2-}$
$$M^{2+} \quad O^{2-} \quad M^{2+} \quad O^{2-} \quad M^{2+}$$

$$M^{2+} \quad O^{2-} \quad M^{2+} \quad O^{2-} \quad M^+$$
b) $\frac{1}{2} O_2 (g) \quad + \quad O^{2-} \quad M^{2+} \qquad M^{2+} \quad O^{2-}$
$$M^+ \quad O^{2-} \quad M^{2+} \quad O^{2-} \quad M^{2+}$$

$$M^{2+} \quad O^{2-} \quad M^{2+} \quad O^{2-} \quad M^{2+}$$
c) $\frac{1}{2} O_2 (g) \quad + \quad O^{2-} \quad M^{2+} \quad \square \quad M^{2+} \quad O^{2-}$
$\quad + M (g) \qquad\qquad O^{2-} \quad M^{2+} \quad O^{2-} \quad M^{2+}$

FIG. 2. The ionization of an oxygen vacancy in MO.

electrons are thermally excited from a lower energy level to a higher one, the associated entropy change being such as to reduce the overall free energy of the system. This second type of ionization is accounted for in the distribution function for the electrons, discussed in Section IV,D.

B. OTHER CRYSTALLINE DEFECTS

The other defects that commonly occur in materials — the dislocation, the grain boundary, and the crystal surface — are also important in considerations of electrical conductivity, for they can interact with point defects and affect both the concentration and the mobility of these defects. In addition, they can become charged as a result of this interaction, and (in the case of mobile dislocations) give a small contribution to the electrical conductivity through their own migration.

The more complex interactions will be treated in Section VI. The ability of dislocations and boundaries to create and destroy lattice sites is briefly described here, since it is important as a mechanism for allowing adjustments in the local point-defect concentrations.

The edge dislocation is a simple example (see Fig. 3); it may be regarded as an extra plane of atoms inserted partway into the crystal between two other complete planes. The edge of the half-plane (the line of the edge dislocation) can act as a source or sink for atoms, which can diffuse from lattice sites in the crystal (leaving vacancies) and be deposited on the edge of the extra plane, or vice versa.

The low-angle grain boundary (tilt) is composed of a set of edge dislocations arranged one above the other and can thus act in a similar fashion. Generally, grain boundaries and crystal surfaces, though not of this simple structure, are regions where atoms can be reversibly attached or detached, thus allowing for variations in the extent of the lattice and in the concentrations of defects.

In compounds, dislocations and surfaces provide a mechanism for creating small concentrations of defects of a single chemical type; this mechanism allows the site ratio in the compound to vary slightly from that of the perfect crystal structure. For example, in a compound M_2O_3 this mechanism allows the ratio of M sites to O sites to vary very slightly from 2/3.

III. ORIGIN OF POINT DEFECTS

A. DEFECTS AS EQUILIBRIUM SPECIES

A notable feature of point defects is that they can occur in thermal equilibrium within the material, a factor which will allow

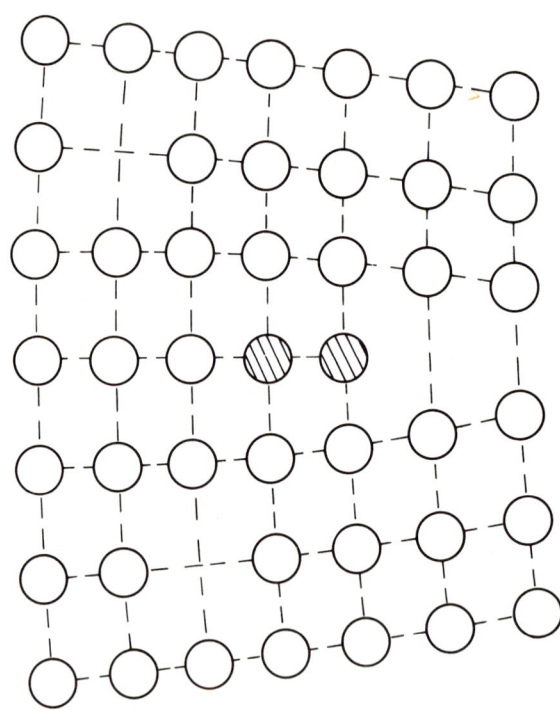

FIG. 3. An edge dislocation. The vacancy concentration can be adjusted by the movement of atoms to or from the edge of the half-plane; addition of the two shaded atoms to the half-plane has created two vacancies.

us to treat them as chemical species in terms of equilibrium constants and the mass-action law. Their occurrence as equilibrium species is illustrated here by using statistical thermodynamics to calculate the concentration of defects that exists in a crystal in equilibrium at a given temperature. The approach was first used by Wagner and Schottky [1].

Under conditions of constant temperature and pressure, the equilibrium structure of a material is that of the lowest free enthalpy (Gibbs free energy) G, where

$$G = H - TS \qquad (2)$$

DEFECT STRUCTURE OF CERAMIC MATERIALS

The introduction of a defect into the crystal will change the free enthalpy of the system by an amount

$$\Delta G = h_d - T \Delta S \qquad (3)$$

where h_d is the enthalpy of formation of the defect and ΔS is the total entropy change caused by the presence of the defect.

The reason for the occurrence of defects as equilibrium species is that despite the fact that they are energetically unfavorable (h_d is positive; the structural energy of the system is increased), the increase they cause in the disorder of the crystal (ΔS is positive) is sufficient to outweigh the other term so that the overall free enthalpy change caused by the introduction of defects can be negative.

The entropy term ΔS contains two contributions:

$$\Delta S = \Delta S_c + s_d \qquad (4)$$

where s_d is the entropy caused by the local distortions around the defect, and ΔS_c is the change in configurational entropy; the latter contribution arises from the disorder represented by the many alternative ways in which the atoms and defects can be deployed on the available sites, and is given by

$$S_c = k \ln \Omega_c \qquad (5)$$

where k is Boltzmann's constant and Ω_c is the number of complexions or distinguishable ways of arranging the atoms and defects on the sites. (A discussion of the entropy contributions can be found in Denbigh's text [2]).

Consider as an example a crystal MX which contains M vacancies and X vacancies as defects. Let the material already contain n_M M vacancies and n_X X vacancies; we can then calculate the free enthalpy change on adding another pair of vacancies (one M vacancy and one X vacancy) to the system. For this addition:

$$\Delta G = h_d - T s_d - T \Delta S_c$$

where h_d and s_d now refer to the enthalpy and entropy of formation of the two added vacancies. From Eq. (5),

$$\Delta G = h_d - Ts_d - kT \ln \frac{\Omega_2}{\Omega_1} \tag{6}$$

where Ω_1 is the number of arrangements of the n_M M vacancies and the n_X X vacancies in the crystal which contains N_M M atoms and N_X X atoms. Since rearrangements of each species (e.g., the M vacancies) among themselves cannot be distinguished,

$$\Omega_1 = \frac{(N_M + n_M)!}{N_M! \; n_M!} \frac{(N_X + n_X)!}{N_X! \; n_X!} \tag{7}$$

For Ω_2 we have one extra pair of vacancies, so

$$\Omega_2 = \frac{(N_M + n_M + 1)!}{N_M! \; (n_M + 1)!} \frac{(N_X + n_X + 1)!}{N_X! \; (n_X + 1)!} \tag{8}$$

Consequently,

$$\frac{\Omega_2}{\Omega_1} = \frac{(N_M + n_M + 1)(N_X + n_X + 1)}{(n_M + 1)(n_X + 1)} \tag{9}$$

Since $N_M, N_X, n_M, n_X \gg 1$, and since for small defect concentrations $N_M > n_M$ and $N_X > n_X$, Eq. (6) becomes

$$\Delta G = h_d - Ts_d + kT \ln \frac{n_M \cdot n_X}{N_M \cdot N_X} \tag{10}$$

For very small defect concentrations, $n_M \cdot n_X$ is small and the last term in Eq. (10) is large and negative, resulting in a lowering of the free enthalpy as defects are introduced. Equilibrium is reached when addition of defects no longer reduces the free enthalpy; at this point $\Delta G = 0$, and from Eq. (10)

$$\frac{n_M \cdot n_X}{N_M \cdot N_X} = \exp \left(\frac{s_d}{k} - \frac{h_d}{kT} \right)$$

$$= \exp \left(-\frac{g_d}{kT} \right) \tag{11}$$

DEFECT STRUCTURE OF CERAMIC MATERIALS 189

This expression shows the number of defects existing in equilibrium at temperature T in terms of g_d, the free enthalpy of formation of the defect. As used above, g_d refers to the formation of the pair of defects, namely the M vacancy and the X vacancy.

In the calculation of the equilibrium defect concentration we have assumed that the location of the M vacancies has no effect on the location of the X vacancies. More specifically, both terms in Ω_1 [see Eq. (7)] assume random placing of the vacancies on the available sites. This is valid so long as interactions between the defects may be neglected, and limits the result to small defect concentrations.

The indication that point defects occur in thermal equilibrium means that in certain respects they may be treated as chemical species in solution in the crystal. For instance, the chemical potential of species A in a solution is given by

$$\mu_A = \left(\frac{\partial G}{\partial n_A}\right)_{T,P,n_B,n_C,\ldots} \tag{12}$$

It represents the change in free enthalpy of the solution caused by an infinitesimal addition of A under conditions of constant temperature, pressure, and quantity of all other species, B, C, etc.

In a parallel fashion we may describe a chemical potential for defects. For example, in the case of the crystal MX considered above, the chemical potential of the M vacancies would be given by

$$\mu = \left(\frac{\partial G}{\partial n_M}\right)_{P,T,N_M,N_X,n_X} \tag{13}$$

which would represent the change in the system free enthalpy caused by the addition of an M vacancy when the temperature, pressure, and numbers of the other species (including defects) were kept constant.

If the crystal consisted of an infinite, dislocation-free array of lattice sites Eq. (13) would be a formal rather than a practical definition. For then,

$$N_M + n_M = N_X + n_X \tag{14}$$

in order to preserve the lattice-site ratio, and the constancy of the other terms in Eq. (13) could not be maintained when n_M was changed. To describe defect potentials where this problem arises the concept of the virtual chemical potential ξ can be introduced in place of the chemical potential.

However, in the real crystalline system there are surfaces, and usually grain boundaries, and dislocations, all of which can act as mechanisms for slightly upsetting the balance of Eq. (14) (see Section II,B). Consequently, the operation represented by Eq. (13) can in principle be performed, and the potential defined in Eq. (13) can then be used. The distinction is not of practical importance, since, as discussed later (Section III,B), reactions involving defects are always written so as to preserve the site ratio, and thus the two notations lead to the same result. In this chapter the definition (13) will be adopted; a discussion of the virtual potential can be found in Kröger's text [3].

The chemical potential per mole [Eq. (12) multiplied by Avogadro's number] is related to the activity a of a species by

$$\mu_A = \mu_A^\circ + RT \ln a_A \tag{15}$$

where μ_A° represents the potential under standard-state conditions (usually the pure material A under 1 atm pressure at temperature T). If ideal solution behavior is obeyed, so that interactions between the members of the various species can be ignored, then the activity is equal to the concentration x, and

$$\mu_A = \mu_A^\circ + RT \ln x_A \tag{16}$$

For the chemical reaction

$$aA + bB \rightarrow cC + dD \tag{17}$$

where A, B, C, and D are the interacting species and a, b, c, d are the respective molar quantities involved in the reaction, Eq. (16) gives for species A

DEFECT STRUCTURE OF CERAMIC MATERIALS 191

$$a\mu_A = a\mu_A^o + RT \ln x_A^a \tag{18}$$

and analogous expressions for the other reactants. The molar free enthalpy change for the overall reaction is

$$\Delta G = c\mu_C + d\mu_D - a\mu_A - b\mu_B \tag{19}$$

and combination with Eq. (18) gives

$$\Delta G = \Delta G^o + RT \ln \frac{x_C^c \cdot x_D^d}{x_A^a \cdot x_B^b} \tag{20}$$

and since $\Delta G = 0$ at equilibrium, we have the following expression for the equilibrium concentrations:

$$\frac{x_C^c \cdot x_D^d}{x_A^a \cdot x_B^b} = \exp\left(-\frac{\Delta G^o}{RT}\right)$$

$$= K$$

$$= \exp\left(\frac{\Delta S^o}{R} - \frac{\Delta H^o}{RT}\right) \tag{21}$$

This is the mass-action law, with K the equilibrium constant and ΔG^o the standard-state free enthalpy change for the reaction (17).

The treatment of defect reactions in solids can follow the same pattern as this treatment of the chemical reaction (17); however, before extending the discussion to defect reactions it is necessary to describe the notation used to represent defects and the guidelines to be followed in the writing of defect reactions.

B. NOTATION FOR DEFECTS AND DEFECT REACTIONS

Because of its great simplicity and elegance, the treatment of defects as individual entities in chemical reactions similar to (17) will be used in the rest of the chapter. The method was initially developed by F. A. Kröger and H. J. Vink [4], and their paper together with the book by Kröger [3] should be consulted for a more extended discussion and for many examples of defect situations.

The method requires a notation in which the representation of defects as chemical species is made clear, and that due to Kröger and Vink will be used throughout this chapter. A number of other notations have been proposed (a convenient summary is given by van Gool [5]) and the literature contains many hybrids of the basic types; however, the notation that follows has won general acceptance and the others will not be considered further.

The defect is represented by a main symbol followed by a subscript and superscript. The main symbol indicates the species concerned, which will be a particular type of atom (represented by its conventional chemical symbol) or a vacancy (shown as V); the subscript shows the site on which the species sits, indicated in terms of the usual occupant of the site in the perfect crystal, or in the case of an interstitial site by $_i$. The superscript shows the effective electrical charge of the defect: this is the difference between the real charge of the defect species and that of the species that would have occupied the site in the perfect crystal. Positive effective charges are shown as dots $^\cdot$, negative effective charges as dashes $'$, and effectively neutral defects by a cross ^{x}. The neutral indication is sometimes omitted.

Use of the notation is best shown by an example. If Al_2O_3 is regarded as an assembly of Al^{3+} and O^{2-} ions in the perfect crystal, then the presence of Mg^{2+}, Cr^{3+}, and Ti^{4+} on Al^{3+} sites would result in Mg'_{Al}, Cr^x_{Al}, and $Ti^\cdot_{Al}$ as defect species. An interstitial Al^{3+} ion is the defect $Al^{\cdots}_i$ and a missing Al^{3+} ion is V'''_{Al}.

The utility of the "effective" representation is that it shows very readily the effect of defects on the overall charge neutrality of the crystal, as well as clearly indicating where coulombic interactions are possible between defects.

By incorporating the notation into chemical equations for defects it is possible to show in a simple and direct fashion the nature and outcome of a defect interaction. However, in writing such equations, it is important just as in conventional equations,

DEFECT STRUCTURE OF CERAMIC MATERIALS   193

   (a) to balance mass on the two sides of the equation
   (b) to balance electrical charge (here, effective charge) on
the two sides of the equation

In addition, because we are dealing with crystalline systems, it is important

   (c) to preserve the site ratio of the crystal

This last requirement recognizes that even when surfaces, grain boundaries, and dislocations are present, equations similar to Eq. (14) must be approximately true if the crystal structure is to be maintained. For example, in MX it is not possible to create M vacancies continually without also adding X vacancies, since the crystal cannot be indefinitely extended by adding M sites only.

Finally we must recognize that the crystal maintains an overall electrical charge neutrality throughout. If defects of a particular effective charge are introduced, there must also be added compensating defects of the opposite charge to achieve this neutrality. Rule (b) above will ensure that neutrality is maintained in a given reaction. The requirement of overall neutrality is expressed in the "neutrality condition":

$$\sum_{\text{negative defects}} [\text{defect charge}] \cdot [\text{defect concentration}]$$

$$= \sum_{\text{positive defects}} [\text{defect charge}] \cdot [\text{defect concentration}]$$

where "negative", "positive", and "charge" are defined in terms of effective charges.

C. INTRINSIC ATOMIC DISORDER

Consider a pure compound $M_m X_x$. If the compound is stoichiometric, so that M and X are present in the proportion m : x, and if the vapor pressures of the two components M and X are small, so that there is little loss of material to the surroundings, then we can only introduce the atomic defect types in certain combinations if

the ratio of M sites to X sites is to be preserved at the value
m : x. These possible combinations are the types of intrinsic
atomic disorder.

There are nine such possible types in all, of which the most
commonly occurring are the following:

Type 1: Schottky disorder, in which corresponding numbers of
vacancies are found on the two sublattices; for the example, this
is achieved by having $[V_M^x]/[V_X^x] = m/x$, where the brackets indicate
concentrations, here expressed as number of vacancies per unit volume;

Type 2: Frenkel disorder, in which equal numbers of vacancies
and interstitials of one of the atom species are found; for instance
if a number of the M atoms leave their normal sites and occupy
interstitial positions, this gives rise to Frenkel disorder on the
M sublattice: for this, $[M_i^x] = [V_M^x]$.

The types of intrinsic atomic disorder are listed in Table 1.
(While the discussion of intrinsic disorder here is in terms of
neutral defects, the disorder types listed in Table 1 also occur in
ionized form, the effective charge on one defect in the pair being
equal and opposite to that on the other. Other forms of Schottky
disorder are, for example,

$$[V_M'] = [V_X^\bullet]$$

$$[V_M''] = [V_X^{\bullet\bullet}]$$

To allow for these cases, no charges are shown in Table 1.)

As an example of Schottky disorder and of the application of
Eq. (21) (the law of mass action) to a defect situation, we may
consider the reaction where equal numbers of V_M^x and V_X^x are introduced into the crystal MX. In the reaction two atoms, M_M^x and X_X^x,
come from the interior of the crystal and sit on the surface, thus
converting the surface atoms they cover into interior atoms.
Labeling the two migrating atoms with an asterisk *, and the surface
sites with subscript s, this is

TABLE 1

Intrinsic Atomic Disorder Types in MX

Name	Description
Schottky disorder	$[V_M] = [V_X]$
Frenkel disorder	$[M_i] = [V_M]$
	$[X_i] = [V_X]$
Interstitial disorder	$[M_i] = [X_i]$
Antistructure disorder	$[M_X] = [X_M]$
	$[M_X] = [V_M]$
	$[M_X] = [X_i]$
	$[X_M] = [V_X]$
	$[X_M] = [M_i]$

$$M_M^{x*} + X_X^{x*} + M_s + X_s \rightarrow M_s^* + X_s^* + M_M^x + X_X^x + V_M^x + V_X^x \qquad (22)$$

Since the asterisks are not physical labels, M_M^{x*} is indistinguishable from M_M^x and Eq. (22) becomes

$$0 \rightarrow V_M^x + V_X^x \qquad (23)$$

which may be compared with Eq. (17). Analogous to Eq. (19) we have

$$\Delta G_S = \mu_{V_M} + \mu_{V_X} \qquad (24)$$

and to Eq. (21) we have

$$[V_M] \cdot [V_X] = \exp\left(-\frac{\Delta G_S^O}{RT}\right)$$

$$= K_S \qquad (25)$$

Here K_S is the equilibrium constant for the creation of Schottky disorder and ΔG_S is the molar free enthalpy change for Eq. (23). Note that the problem is the same as that treated by statistical methods in Section III,A; when the concentrations in Eq. (25) are written in terms of site fractions (the ratio of the number of defects at a particular type of site to the total number of sites of that same type in the material), Eq. (25) is identical with Eq. (11), remembering the approximations following Eq. (9).

Since for Schottky disorder $[V_M^x] = [V_X^x]$, we have

$$[V_M^x] = [V_X^x] = \exp\left(-\frac{\Delta G_S^0}{2RT}\right) \tag{26}$$

as the expression for the individual concentrations.

As a second example, the equation for Frenkel disorder on the M sublattice in MX is

$$M_M^x + V_i^x \rightarrow M_i^x + V_M^x \tag{27}$$

Applying the mass-action law,

$$\frac{[M_i^x] \cdot [V_M^x]}{[M_M^x] \cdot [V_i^x]} = [M_i^x] \cdot [V_M^x]$$

$$= K_F$$

$$= \exp\left(-\frac{\Delta G_F^0}{RT}\right) \tag{28}$$

where for Frenkel disorder $[M_i^x] = [V_M^x]$. The terms $[M_M^x]$ and $[V_i^x]$, which represent the host lattice constituents, are virtually constant for small defect concentrations; they can therefore be put at unity (when concentrations are expressed as site fractions), or they can be put as a constant and taken into K_F, when other concentration units are used. It must be remembered that the values of constants such as K_F depend as in the above case on the units used for concentration.

To emphasize the convenience of the mass-action approach we can treat the same defect situation by the statistical method.

DEFECT STRUCTURE OF CERAMIC MATERIALS   197

Since the number of M sites is unchanged in Frenkel disorder,

$$\Omega_1 = \frac{N_M!}{n_M!\,(N_M - n_M)!} \frac{N_i!}{n_i!\,(N_i - n_i)!} \tag{29}$$

and

$$\frac{\Omega_2}{\Omega_1} = \frac{(N_M - n_M)(N_i - n_i)}{(n_M + 1)(n_i + 1)} \tag{30}$$

giving

$$\frac{n_i \cdot n_M}{N_i \cdot N_M} = \exp\left(-\frac{g_F}{kT}\right) \tag{31}$$

where n_i is the number of interstitial M atoms, N_i the number of interstitial sites, and g_F the free enthalpy for the formation of one vacancy and one interstitial defect.

D. EXTRINSIC ATOMIC DISORDER

By extrinsic atomic disorder we mean any type of defect structure other than those listed in Table 1 and discussed in Section III,C; the disorder types described there are not dependent for their existence on the presence of impurities or on a departure from the stoichiometric composition of the sample, but are features of a pure, stoichiometric crystal.

From this definition, there are two main types of extrinsic atomic disorder: in one the defect structure is dictated by the presence of impurities, and in the other the crystal has adjusted to variations in the partial pressure of one of its components in the surrounding atmosphere, with resulting departure from stoichiometric composition. This definition is not universally followed and care should be taken to identify the usage when encountering the terms in the literature.

1. Nonstoichiometry

Under conditions where a crystal is in equilibrium with the partial pressure of one of its components in an atmosphere above it, adjustments of the partial pressure will alter the composition of the crystal; for a binary compound, use of the phase rule shows that such adjustments will completely determine the composition.

Taking as example a binary oxide MO under equilibrium conditions, so that the chemical potential of oxygen in the gas phase and in the crystal are equal, we are saying that variations in oxygen pressure, $p(O_2)$, will cause variations in the ratio M : O in the solid.

Let $p(O_2)$ be fixed at the value $p(O_2)'$ where the oxide is stoichiometric, so that $[M]_{tot.} = [O]_{tot.}$. Here $[M]_{tot.}$ is the overall M concentration to which there may be several contributing terms, such as $[M_M^x]$, $[M_i^x]$, and $[M_O^x]$. If $p(O_2)$ is then increased, $[O]_{tot.}$ is increased relative to $[M]_{tot.}$. The excess oxygen is incorporated by defect reactions; some possible mechanisms are

$$V_i^x + \tfrac{1}{2}O_{2(g)} \rightarrow O_i^x \tag{32}$$

$$\tfrac{1}{2}O_{2(g)} \rightarrow O_O^x + V_M^x \tag{33}$$

where O_O indicates the incorporation of an oxygen ion in an oxygen site, and for the moment it is assumed that the defects are not ionized. Using the mass-action law it is immediately seen that the defect concentrations depend specifically on the environmental oxygen pressure:

$$[O_i^x] = K_1\, p(O_2)^{1/2} \tag{34}$$

or

$$[V_M^x] = K_2\, p(O_2)^{1/2} \tag{35}$$

In writing these equations, it has been assumed that the defect concentrations $[O_i^x]$ and $[V_M^x]$ are small, and hence that the concentrations $[V_i^x]$ and $[O_O^x]$ of the normal lattice constituents are, as noted earlier, essentially constant. The situation where the assumption is incorrect owing to the presence of large defect concentrations is discussed in Section V,B.

DEFECT STRUCTURE OF CERAMIC MATERIALS

If $p(O_2)$ falls below $p(O_2)'$, an oxygen deficit occurs in the crystal. This deficit may, for example, be achieved by either of the following mechanisms:

$$O_O^x \rightarrow \tfrac{1}{2}O_{2(g)} + V_O^x \tag{36}$$

$$V_i^x + M_M^x + O_O^x \rightarrow M_i^x + \tfrac{1}{2}O_{2(g)} \tag{37}$$

The defect concentrations are then

$$[V_O^x] = K_3\, p(O_2)^{-1/2} \tag{38}$$

$$[M_i^x] = K_4\, p(O_2)^{-1/2} \tag{39}$$

By varying the pressure of the more volatile component [usually $p(O_2)$] over the system, the crystal composition should be variable (within the experimental limits of the apparatus) over the compositional range of stability of the oxide. The width of this range and whether it includes the stoichiometric composition depend on the particular system; some examples are shown in Table 2.

2. Impurity

The existence of an impurity L in solution in a crystal MO gives rise to point defects which may be substitutional (L_M^x or L_O^x) or interstitial (L_i^x) or both. These defects may also occur in ionized form.

The impurity may occur in quasi-equilibrium or complete equilibrium. In the former case, the impurity enters the crystal at some early stage in its preparation, and continues to exist in the crystal even in the absence of a partial pressure of the impurity species in the surrounding atmosphere; this can arise where the impurity is a species having low volatility. The crystal then contains a fixed total quantity of the impurity which may be divided into a number of different defect types, e.g.,

$$[L]_{tot.} = \text{const.} = [L_i^x] + [L_M^x] \tag{40}$$

TABLE 2

Homogeneity Range and Oxygen Pressure Range of 3d Oxides[a]

Oxide	Stability range[b]			p(O) range[c]		
	$x_{min.}$	$x_{max.}$	Δx	Max	Min	$\Delta p(O)$
TiO	0.80	1.30	0.50	44.1	41.5	2.6
Ti_2O_3	1.501	1.512	0.011	41.5	30.1	11.4
TiO_2	1.992	2.000	0.008	25.7	--	--
VO	0.80	1.30	0.50	34.5	33.2	1.3
MnO	1.000	1.18	0.18	34.7	10.7	24.0
FeO	1.045	1.200	0.155	20.5	19.2	1.3
Fe_3O_4	1.336	1.381	0.045	19.2	14.0	5.2
CoO	1.000	1.012	0.012	14.5	2.5	12.0
NiO	1.000	1.001	0.001	16.5	--	--
Cu_2O	0.500	0.5016	0.0016	9.6	7.0	2.6

[a] This table is reprinted from Ref. 6, p. 22, by courtesy of the North-Holland Publishing Company.

[b] The stability range is expressed in terms of x in MO_x.

[c] $p(O) = -\log_{10} p(O_2)$

In the complete equilibrium situation, a partial pressure of the impurity in the atmosphere is maintained in equilibrium with the concentration of impurity in the crystal. We may then write possible reactions for the incorporation of the impurity in a manner similar to Eqs. (32) and (33):

$$L_{(g)} \rightarrow L_O^x + V_M^x \qquad (41)$$

$$L_{(g)} \rightarrow L_M^x + V_O^x \qquad (42)$$

$$L_{(g)} \rightarrow L_i^x \qquad (43)$$

The mass-action law may then be applied to these equations.

E. RADIATION DAMAGE AND PLASTIC DEFORMATION

In addition to creation by thermal equilibration, there are two mechanical methods for creating point defects in a crystal. One occurs in plastic deformation and is the result of the intersection of jagged dislocations — depending on the Burgers vectors of the dislocations, a string of vacancies or of interstitials is left in the wake of the intersection. The second occurs as a result of exposure to radiation, e.g., neutron bombardment; it is part of the process of radiation damage, and has been used to provide considerable insight into the structure and behavior of individual defects.

Discussion of these mechanisms for defect generation may be found in the referenced books by Nabarro [7] and Chadderton [8].

IV. DEFECTS AND ELECTRON ENERGY LEVELS

A. BAND STRUCTURE AND DEFECT LEVELS

In the preceding discussion we have mentioned ionization of defects and the need for overall electrical neutrality, but have otherwise treated defects which have had zero effective charge. In discussing electrical conductivity we shall be interested only in defects which can act as charge carriers, and in the convention adopted here such defects must be effectively charged.

This may not seem obvious when, for instance, in an ionic oxide, the effectively neutral oxygen ion, O_O^x, carries a real charge, O^{2-}, so that its movement would imply conduction. However, as O_O^x moves one way, a vacancy moves the other. Where the vacancy carries no real charge, the defect is $V_O^{\cdot\cdot}$ which, being effectively charged, contributes to the conductivity in the present convention; where the vacancy is effectively neutral, V_O^x, it carries a real charge of 2- (from two localized electrons) so that exchange with the oxygen ion has O^{2-} moving one way and two electrons localized at a vacancy moving the other with no resultant charge transfer. Consistent results are achieved either by working with real charges or by working with effective charges, and for the reasons outlined earlier, the latter convention is adopted here.

Because of this concern with effectively charged defects, electronic defects and the ionization of atomic defects are discussed more fully in this section. Band structure and electronic conduction is the subject of Chapter 4; here we note some particular features needed for a description of the interactions between atomic defects and electronic carriers.

In a single atom, the possible states for electrons occur at distinct and separate energies (energy levels); as the atoms are brought together to form the solid, those electron states which do not overlap with the neighboring atoms (the "inner" electrons) remain as narrow local levels, and therefore electrons in these states are localized around particular atoms in the solid. The outer electron states of the individual atoms, which do overlap with their neighbors, are transformed into a set of new states in the solid, which taken together cover a band of energy values (see Fig. 4).

As seen in Fig. 4, the spacing between the atoms, or better the extent of overlap of the electron states on the individual atoms, determines the width of the energy band; this width is the range of energies covered by the electron states in the solid contributed by the one particular level of the free atoms. The overlap also influences the size of the energy gaps, which are the ranges of energy for which there are no electron states in the solid and which are as a consequence inaccessible to electrons.

The outer occupied level in the atoms (containing the valence electrons) forms the valence band in the solid; the lowest unoccupied level in the atom (its occupation would occur in an excited state of the atom) forms the conduction band in the solid. The gap between the two is the band gap. In insulating and semiconducting

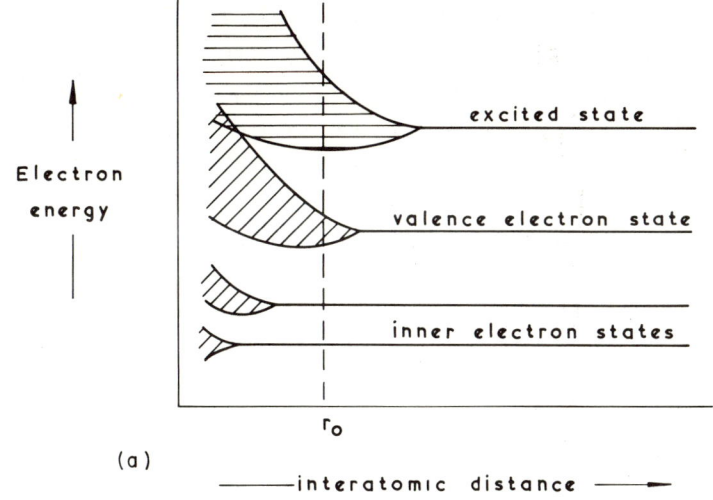

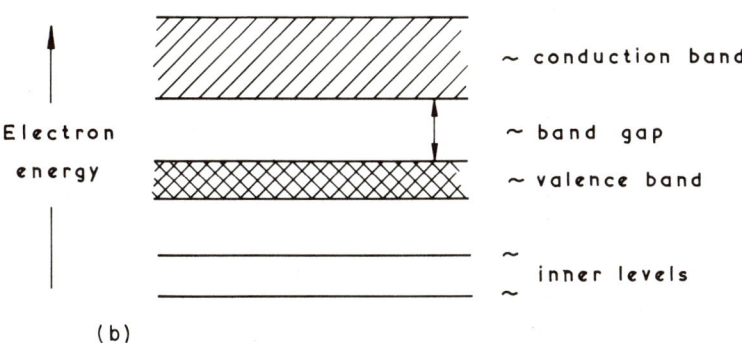

FIG. 4. (a) Electron energy states in an elemental solid as a function of interatomic spacing. (b) The electron energy states at the equilibrium interatomic spacing r_0 are plotted against a general space coordinate in the solid to give a schematic energy-band diagram.

materials at low temperature the valence band is fully occupied by electrons and the conduction band is empty. Consequently, unless electrons are excited across the band gap there are no new energy states electrons can occupy on application of a field; the field thus can have no effect on the electron energies or momenta and the crystal is nonconducting.

Electronic charge carriers are provided by electrons in the conduction band or by missing electrons (electron holes) in the valence band; depending on the width of the band in which these carriers move, two conduction types can be distinguished (see Chapter 4). In broad bands the electrons are delocalized and move through the crystal as free electrons subjected to a small perturbing potential provided by the atoms on their sites. In narrow bands the electrons are localized around particular atoms and move by jumping from the potential well around one atom to that around another. The electron mobilities for the two types are very different (see Section VIII,C and Chapter 4).

When a defect is added to the crystal it destroys the periodicity and, as a result, introduces local electron states that are different from the host states and which may lie within the band gap; whether these states are occupied by electrons or not depends on the particular defect, as discussed below.

If the defect, placed as an effectively neutral species in the crystal, provides an occupied level within the band gap, then it is termed a donor, since the electron occupying the level can be excited to a state in the conduction band (donated); if the neutral defect provides an unoccupied level in the gap, then it is termed an acceptor, since an electron may be excited from a state in the valence band of the host into the defect level (accepted).

These considerations may be illustrated by an example, for which it is convenient to choose an ionic compound; first because many of the ceramic materials of interest are ionic, and second because, since the electrons are predominantly associated with particular ions, the electron band that an electron occupies can be described in terms of the ion to which the electron is attached. We consider accordingly an ionic oxide MO.

For the oxygen ions O^{2-} the inner electrons occupy distinct local levels in the solid, whereas the outer electrons, by partially interacting with other O^{2-} ions, give rise to an electron band. The energy of the highest occupied band arising from the oxygen-ion levels is given by the energy of the reaction

$$O^- \text{ (lattice)} + e^- \text{ (free space)} \to O^{2-} \text{ (lattice)} \qquad (44)$$

which is the energy needed to attach the outermost electron to the oxygen ions in the solid. The energy for the reaction (44) contains three main terms:

(a) the electron affinity of the O^- ion, which is E_A, the energy of the reaction

$$O^- \text{ (free space)} + e^- \text{ (free space)} \to O^{2-} \text{ (free space)} \qquad (45)$$

(b) the electrostatic energy E_M gained by adding an electron into the positive potential region in the unrelaxed crystal

$$e^- \text{ (free space)} \to e^- \text{ (vacant anion site)} \qquad (46)$$

(c) local ion movements (polarization) following the reaction (46), which reduce the electrostatic energy contribution by an amount E_P.

In sum, the energy of the band measured with respect to an electron in free space is

$$E_{O^{2-}} = E_A - E_M + E_P \qquad (47)$$

The convention is to label a band or level with the species that is formed when the level is occupied; $E_{O^{2-}}$ thus gives the approximate energy of the O^{2-} band.

Similarly the outer electrons of the cation M^{2+} will give rise to bands. The energy of the highest occupied band arising from the cation levels is given by

$$M^{3+} \text{ (lattice)} + e^- \text{ (free space)} \to M^{2+} \text{ (lattice)} \qquad (48)$$

which is the energy needed to attach the outermost electron to the cation in the solid. For a nontransition-metal ion, this is

$$E_{M^{2+}} = -E_I + E_M + E_P \qquad (49)$$

when E_I is the third ionization potential of the metal M, and where

E_M enters with changed sign since the electron is now added to a region of negative potential (a cation site) within the crystal.

For transition-metal ions Eq. (49) contains an added term to allow for the fact that the field provided by the surrounding oxygen ions stabilizes certain d orbitals at the expense of others and hence makes certain ionization states favorable. This aspect can be studied further in Orgel's text [9].

The valence band in MO is the highest occupied band, and since the O^{2-} and M^{2+} bands are the highest occupied bands contributed by the constituents of the oxide, the nature of the valence band depends on the relative values of $E_{O^{2-}}$ and $E_{M^{2+}}$; diagrams for MgO and NiO are shown in Fig. 5, where it can be seen that the valence bands are respectively the Ni^{2+} band and the O^{2-} band. This situation is representative of transition-metal oxides and nontransition-metal oxides, respectively.

The conduction band in MO is the lowest empty band and could in principle be either the next higher oxygen band or the next higher cation band (O^{3-} or M^+, respectively, since the band is conventionally labeled by the species formed by putting an electron into it). These have energies given by

O^{2-} (lattice) + e^- (free space) → O^{3-} (lattice)

M^{2+} (lattice) + e^- (free space) → M^+ (lattice) (50)

The electron affinity of O^{2-} is large and positive, so $E_{M^+} < E_{O^{3-}}$, and the M^+ band forms the conduction band (in Fig. 5, Ni^+ and Mg^+).

An essentially similar technique can be applied to the location of a defect level within the band gap provided the electron is strongly localized around the defect. For instance, the oxygen vacancy in the effectively neutral condition (atom vacancy) consists of two electrons localized at the ion vacancy site; their electrostatic environment (the rest of the crystal) is similar to that of the electrons on normal oxygen ions except that there is no "chemical" binding to an oxygen atom at the site. They therefore will occupy

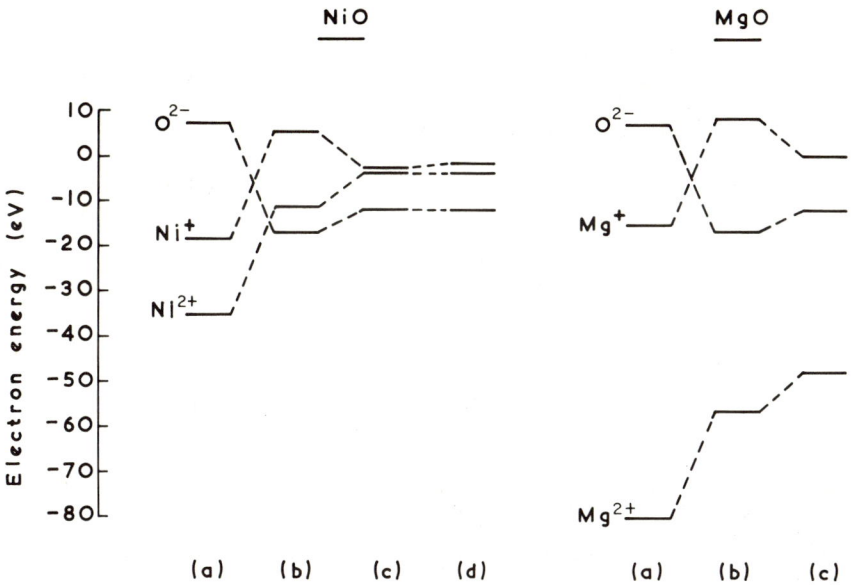

FIG. 5. Energy-band diagrams for NiO and MgO: (a) Free-ion levels (given by E_A and E_I) [19]; (b) ions in the unrelaxed crystal (includes E_M); (c) ions in the polarized crystal (includes E_p) [20]; (d) correction for crystal-field stabilization.

higher energy levels than electrons in the local oxygen-ion levels. In short, V_O^x acts as a donor. Similarly, electrons attached to interstitial M atoms occupy a region of high potential (they are surrounded by anions), and M_i^x acts as a donor.

Reciprocal arguments indicate that V_M^x and O_i^x are able to act as acceptors; the final band picture for a nontransition-metal oxide is then as shown in Fig. 6.

Again, all levels and bands are labeled with the species formed by occupation of the level or band. It must be emphasized that this is a convention only; for example, the level V_M' indicates that the crystal can contain either V_M^x or V_M' as defects depending on whether the level is empty or filled, respectively. Similarly, $V_O^{\cdot}$ indicates $V_O^{\cdot\cdot}$ and $V_O^{\cdot}$ as potential defects in the crystal. A level indicates

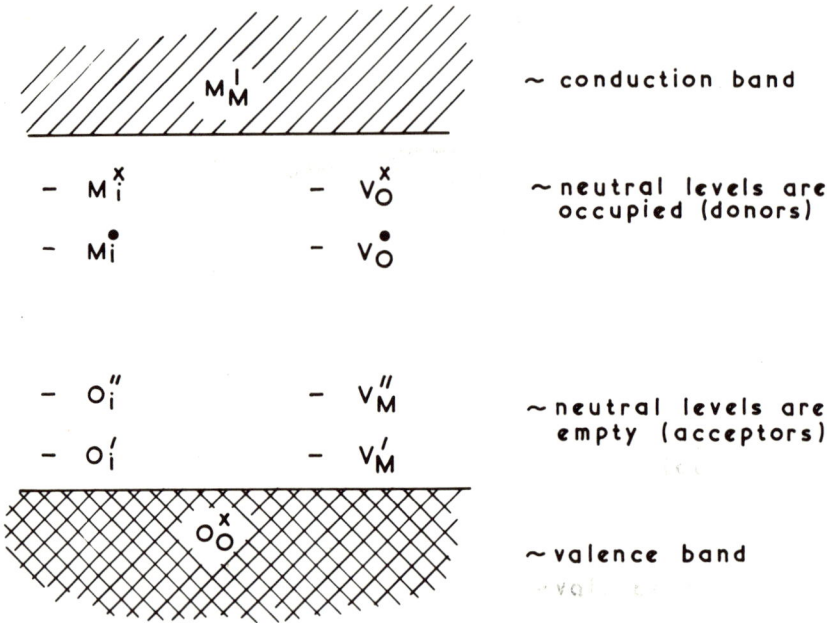

FIG. 6. Electron energy-band diagram for the nontransition-metal oxide MO, including electron energy levels for some possible native point defects.

two possible defects depending on its occupancy; for labeling purposes only one of the two (that corresponding to the filled condition) is used. In Fig. 6 it is assumed that the defects can be doubly ionized; e.g., the defect V_M^x can accept first one electron, giving V_M', which can in turn accept a second electron, giving V_M''. Addition of the second is obviously more difficult than that of the first, which is accounted for in the sequence of the levels in Fig. 6.

Impurity levels can also be placed on a diagram such as Fig. 6. Substitutional impurities generally act as donors or acceptors depending on whether they have more or fewer valence electrons than the species they replace.

For broad-band conduction where the electrons are not localized to particular atoms in the solid, the concepts of valence band and

DEFECT STRUCTURE OF CERAMIC MATERIALS   209

conduction band of course still apply, but their identification with particular ion types becomes less valid. Also, the detailed calculation of defect energy levels is altered, since the electron associated with the defect is not to be regarded as "chemically" bound to the center, but rather as interacting more distantly with the defect by Coulombic forces.

Further discussion of band structures in oxides can be found in Kröger's paper [10] and in the recent review by Goodenough [11].

B. INTRINSIC ELECTRONIC DISORDER

At temperatures above $0°K$ it is possible for electrons occupying low levels to be thermally excited into higher empty levels. In particular they can be excited from levels at the top of the valence band into levels at the bottom of the conduction band. Each excitation gives rise to a free electron e' in the conduction band, and a free electron hole $h^{\cdot}$ in the valence band. These act as negative and positive charge carriers, respectively.

Using our chemical representation for defect reactions, and since the numbers of electrons in the valence band and of unoccupied states in the conduction band are virtually unchanged, we may for small defect concentrations write this excitation process as

$$e_{\substack{\text{valence} \\ \text{band}}} + h_{\substack{\text{conduction} \\ \text{band}}} = e'_{\substack{\text{conduction} \\ \text{band}}} + h^{\cdot}_{\substack{\text{valence} \\ \text{band}}}$$
$$= e' + h^{\cdot} \qquad (51)$$

from which by the mass-action law

$$[e'] \cdot [h^{\cdot}] = n \cdot p = K_i \qquad (52)$$

where

$$K_i = C \exp\left(-\frac{\Delta E_i}{RT}\right) \qquad (53)$$

ΔE_i is the band-gap width; for C, consult Chapter 4.

(It may be noted that in writing defect equations involving electrons and holes, the mass balance requirement of the equation is usually neglected with respect to the electronic defects; the electronic species are assumed to have negligible mass, so that e^x and h^x in Eq. (51) and other equations where they occur can be dropped. Then Eq. (51) can be rewritten

$$0 \rightarrow e' + h^{\cdot} \tag{51a}$$

which leads to Eq. (52) as before.

Another example of this usage is in Eq. (70) below, where the term

$$2e^x_{\text{valence band}}$$

has been dropped from the left-hand side.)

If the process (51) dominates other electron and hole formation processes, and the concentration of charged atomic defects is small, then charge neutrality may be achieved by equal concentrations of electrons and holes

$$n = p \tag{54}$$

which is the condition for intrinsic electronic disorder.

C. EXTRINSIC ELECTRONIC DISORDER

If electrons or holes are generated predominantly by the ionization of defects, then the electronic carriers are extrinsic.

Electrons arise by excitation from donor defects, e.g., from Fig. 6:

$$M_i^x \rightarrow e' + M_i^{\cdot} \;;\quad \frac{[M_i^{\cdot}] \cdot n}{[M_i^x]} = K_{M_i^x} \tag{55}$$

$$M_i^{\cdot} \rightarrow e' + M_i^{\cdot\cdot} \;;\quad \frac{[M_i^{\cdot\cdot}] \cdot n}{[M_i^{\cdot}]} = K_{M_i^{\cdot}} \tag{56}$$

$$V_O^x \rightarrow e' + V_O^{\cdot} \;;\quad \frac{[V_O^{\cdot}] \cdot n}{[V_O^x]} = K_{V_O^x} \tag{57}$$

DEFECT STRUCTURE OF CERAMIC MATERIALS

$$V_O^{\bullet} \rightarrow e' + V_O^{\bullet\bullet} \;;\quad \frac{[V_O^{\bullet\bullet}] \cdot n}{[V_O^{\bullet}]} = K_{V_O^{\bullet}} \tag{58}$$

Holes arise by excitation of electrons from the valence band into acceptor levels:

$$O_i^x \rightarrow h^{\bullet} + O_i' \;;\quad \frac{[O_i'] \cdot p}{[O_i^x]} = K_{O_i'} \tag{59}$$

$$O_i' \rightarrow h^{\bullet} + O_i'' \;;\quad \frac{[O_i''] \cdot p}{[O_i']} = K_{O_i''} \tag{60}$$

$$V_M^x \rightarrow h^{\bullet} + V_M' \;;\quad \frac{[V_M'] \cdot p}{[V_M^x]} = K_{V_M'} \tag{61}$$

$$V_M' \rightarrow h^{\bullet} + V_M'' \;;\quad \frac{[V_M''] \cdot p}{[V_M']} = K_{V_M''} \tag{62}$$

The various equilibrium constants are given by expressions of the type

$$K_x = K_x^o \exp\left(-\frac{\Delta E_x}{RT}\right) \tag{63}$$

where ΔE_x is the energy separation between the defect level and the energy band involved in the excitation. Impurity levels can ionize similarly. In all the situations where any of Eqs. (55)-(62) produce the majority of the electronic carriers, the carriers are extrinsic; the product n·p, as always, is constant at a given temperature, but $n \neq p$.

D. OCCUPATION OF LEVELS

In the preceding sections the location of electron states on an energy scale has been described and some of the possible electron transitions from one state to another have been noted. Whether a given level is occupied or not depends not only on its location in the band diagram but also on the Fermi level E_f in the material.

From Fermi statistics the probability that a given level of energy E is occupied is given by

$$P(E) = \frac{1}{1 + \exp[(E - E_f)/kt]} \quad (64)$$

For $E = E_f$, $P(E) = \frac{1}{2}$.

If at some finite temperature the crystal and its defects are assembled, with the exception of the electrons, and are then gradually fed in, the levels are filled according to Eq. (64). As shown in Fig. 7, the lowest levels are fully occupied, half the available levels at the same energy as E_f are occupied, and the highest levels are empty. Levels below E_f are progressively more occupied, and levels above it progressively less.

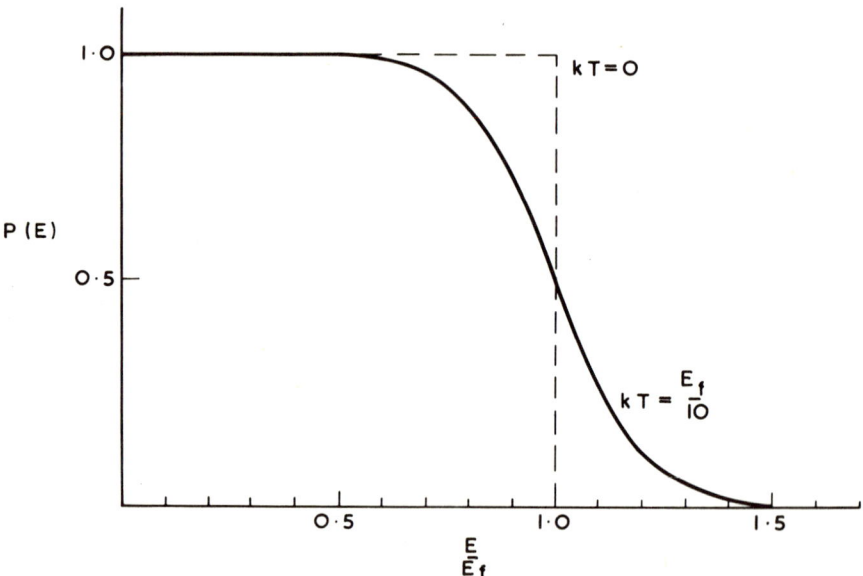

FIG. 7. The Fermi-Dirac distribution function; the probability P(E) that an electron state of energy E will be occupied is shown for two temperatures. Note that the energy span of partly occupied states is of the order kT.

The position of the Fermi level on the band diagram depends, in this picture, on how many electrons are available for filling the levels. In general, reduction processes (removal of oxygen atoms from the crystal, leaving electrons behind) increase the electron concentration and raise the Fermi level; correspondingly, oxidation processes lower the Fermi level.

More quantitatively, the Fermi level is related to the concentration n of free electrons. If there are N_c available states for electrons at the bottom of the conduction band E_c, then

$$\frac{n}{N_c} = P(E_c) = \frac{1}{1 + \exp[(E_c - E_f)/kT]} \tag{65}$$

for $n \ll N_c$, this gives

$$\frac{n}{N_c} = \exp\left(\frac{E_f - E_c}{kT}\right) \tag{66}$$

whence

$$E_f - E_c = kT \ln \frac{n}{N_c} \tag{67}$$

This indicates, as discussed qualitatively above, that increasing n, the free electron concentration, raises the Fermi level.

V. DEFECT EQUILIBRIA IN CRYSTALS

A. SMALL DEFECT CONCENTRATIONS

In the earlier sections, we have described the various types of point defect that can occur in a crystal; we shall now see how the concentrations of these defects are interrelated, and how they change with changes of sample temperature, with nonstoichiometry, and with impurity level. Initially we shall assume small concentrations of the defects so that activities may be replaced by concentrations in the mass-action equations. The discussion is given in terms of the oxide MO but is generally applicable to binary compounds.

1. Effects of Oxygen-Pressure Variation

In the case of the general binary compound, we can vary the composition of the compound by varying the pressure of the more volatile component kept in equilibrium with the solid (see Section II,D,1): for MO this is usually the oxygen pressure $p(O_2)$.

Consider MO to contain, as defects, M vacancies, O vacancies, electrons, and holes, and take as the energy-level diagram the situation shown in Fig. 8, where the vacancy levels are taken, for simplicity in the opening discussion, to lie very close to or within the conduction and valence bands. If MO is an insulator or semiconductor the Fermi level lies within the band gap and the vacancies occur mainly in the fully ionized forms V_M'' and $V_O^{\cdot\cdot}$. (The highest occupied V_M level is V_M'' and the lowest empty V_O level is $V_O^{\cdot}$, so that defects present are V_M'' and $V_O^{\cdot\cdot}$).

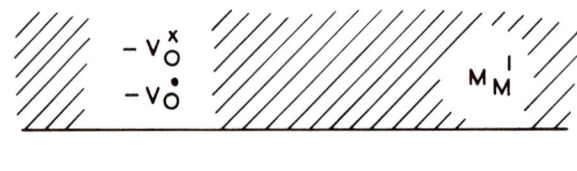

 E_f

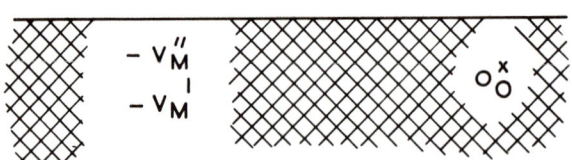

FIG. 8. Hypothetical energy-band diagram for MO.

For the crystal considered, there are four unknown defect concentrations: n, p, $[V_M'']$, and $[V_O^{\cdot\cdot}]$. We may see how these concentrations vary with changes in oxygen pressure over the crystal, assuming equilibrium between the solid and the gas phase. There are four independent relations between the unknown concentrations:

(a) that for creation of ionic disorder

$$0 \rightarrow V_M'' + V_O^{\cdot\cdot} \; ; \qquad [V_M''] \cdot [V_O^{\cdot\cdot}] = K_S \qquad (68)$$

(b) that for creation of electronic disorder

$$0 \rightarrow e' + h^{\cdot} \; ; \qquad np = K_i \qquad (69)$$

(c) any reaction expressing incorporation of oxygen into the solid in terms of the defects present, e.g.,

$$\tfrac{1}{2} O_{2(g)} \rightarrow O_O^x + V_M'' + 2h^{\cdot} \; ;$$

$$[V_M''] p^2 = K_A p(O_2)^{1/2} \qquad (70)$$

(d) the condition for electrical neutrality

$$2[V_M''] + n = 2[V_O^{\cdot\cdot}] + p \qquad (71)$$

It should be noted with respect to Eqs. (68) and (69) that there is no assumption that the crystal is exhibiting one of the intrinsic disorder types, which would require, in addition, either $[V_M''] = [V_O^{\cdot\cdot}]$ or $n = p$. What the equations recognize is that the mass-action expressions, derived from consideration of these disorder types, are always valid and may therefore be used as convenient general relations between the defects present in the crystal.

With respect to Eq. (70), another possible reaction to express oxygen incorporation is

$$V_O^{\cdot\cdot} + 2e' + \tfrac{1}{2} O_{2(g)} \rightarrow O_O^x$$

for which the mass-action expression is

$$K = ([V_O^{\cdot\cdot}] n^2 p(O_2)^{1/2})^{-1}$$

Use of Eqs. (68) and (69) shows that this is the same information as given by Eq. (70); in general, any reaction describing incorporation of oxygen into the solid is equally suitable, provided it is written in terms of defects known to be present.

Solution of the four equations (68)-(71) in terms of values of the defect concentrations at a given oxygen pressure is complicated by the fact that the first three are written as products of concentrations whereas the last is written as a sum of concentrations. Accordingly, an approximate method proposed by Brouwer [12] and largely extended by Kröger and Vink [4] is commonly adopted.

In this method, the concentrations are presented in the form of a Brouwer diagram; by taking logarithms of both sides in the first three equations we find linear relations involving the logarithms of the defect concentrations and of the oxygen pressure. To obtain a similar linear relation from Eq. (71) we make the assumption that on each side of the neutrality equation one of the concentrations is so dominant as to make the other negligible.

The validity of this approximation will be considered below; for the present, continuing with the example, and considering a situation where the concentrations of defects are such that $2[V_M''] \gg n$ and $p \gg 2[V_O^{\cdot\cdot}]$, Eq. (71) becomes the approximation

$$2[V_M''] = p \tag{72}$$

This expression can be written as a linear relation between logarithms of concentrations, so that there are now four such relations with four unknown defect concentrations.

For the simpler situations, the dependence of the concentrations on the oxygen pressure can usually be worked directly from the mass-action equations and the neutrality approximation, Eq. (72). In the present instance, $[V_M''] p^2 = K_A \, p(O_2)^{1/2}$ but $2[V_M''] = p$. Substituting,

$$4[V_M'']^3 = \frac{p^3}{2} = K_A \, p(O_2)^{1/2} \tag{73}$$

whence

$$2[V_M''] = p = K_A' \, p(O_2)^{1/6} \tag{74}$$

DEFECT STRUCTURE OF CERAMIC MATERIALS 217

Since $np = K_i$,

$$n = \frac{K_i}{K_A'} p(O_2)^{-1/6} \qquad (75)$$

and since $[V_M''][V_O^{\cdot\cdot}] = K_S$,

$$[V_O^{\cdot\cdot}] = \frac{2K_S}{K_A'} p(O_2)^{-1/6} \qquad (76)$$

The four concentrations are then known as a function of oxygen pressure provided Eq. (72) is a valid approximation. The results (74), (75), and (76) can be conveniently presented on a graph of log [defect] against log $p(O_2)^{1/2}$, where they appear as straight lines; they are shown on the right-hand sides of the plots in Fig. 9.

We have two possibilities for the relative magnitudes of $[V_O^{\cdot\cdot}]$ and n, depending on the nature of the crystal and particularly on the relative energies of Eqs. (68) and (69). If the creation of ionic disorder is favored relative to that of electronic disorder, as would occur in a wide band-gap material, then $K_S > K_i$, and from Eqs. (75) and (76), $[V_O^{\cdot\cdot}] > n$. This is shown in Fig. 9(a). Correspondingly, where electronic disorder is favored, so that $K_i > 2K_S$, one finds $n > [V_O^{\cdot\cdot}]$, as shown in Fig. 9(b).

Considering Fig. 9(a), it is seen that as the oxygen pressure is lowered, p falls and $[V_O^{\cdot\cdot}]$ rises, so that the approximation [Eq. (72)] for the neutrality condition [Eq. (71)] becomes invalid. From Fig. 9(a) it can be seen that a better expression contains three terms,

$$2[V_M''] = 2[V_O^{\cdot\cdot}] + p \qquad (77)$$

Substituting into Eq. (77) from Eqs. (68) and (70), full expressions for the three defect concentrations are then

$$p^3 (1 + 2 \frac{K_S}{K_A} \frac{p}{p(O_2)^{1/2}}) = 2K_A p(O_2)^{1/2} \qquad (78)$$

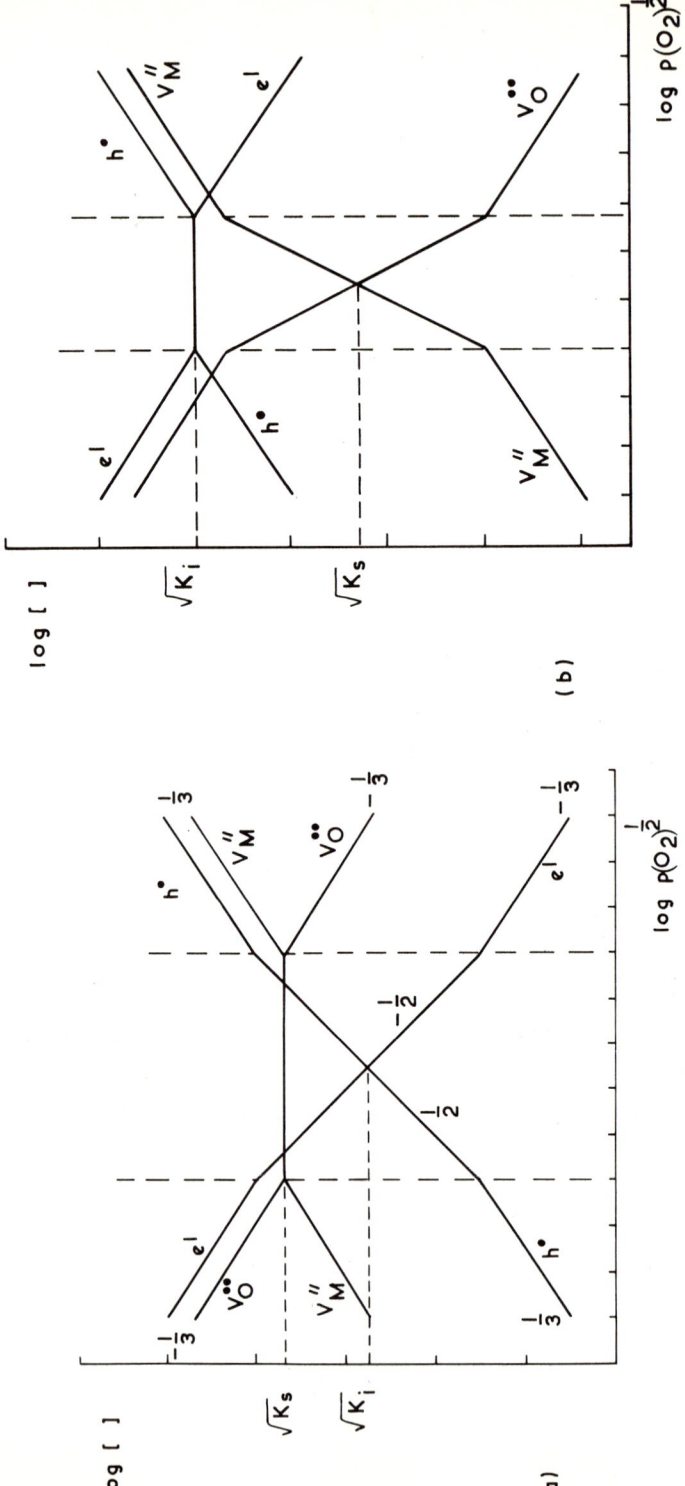

FIG. 9. Defect concentrations in the oxide MO (see Fig. 8). The gradients of the various defect lines are as indicated. Note that horizontal scale = $\frac{1}{2}$ vertical scale to emphasize gradient variation. (a) $K_S > K_i$. Neutrality approximation ranges from left to right are $n = 2[V_O^{\cdot\cdot}]$, $[V_O^{\cdot\cdot}] = [V_M'']$, and $p = 2[V_M'']$. (b) $K_S < K_i$. Neutrality approximation ranges from left to right are $n = 2[V_O^{\cdot\cdot}]$, $n = p$, and $p = 2[V_M'']$.

DEFECT STRUCTURE OF CERAMIC MATERIALS 219

$$2[V_M'']^2 = \sqrt{K_A[V_M'']} \cdot p(O_2)^{1/4} + 2K_S \tag{79}$$

$$2[V_O^{\cdot\cdot}]^2 = 2K_S - \sqrt{K_A/K_S\,[V_O^{\cdot\cdot}]^3} \cdot p(O_2)^{1/4} \tag{80}$$

At high $p(O_2)$ the dependences of the defect concentrations on $p(O_2)$ given by these expressions are identical with those found directly using the approximation (72); slopes 1/3, 1/3, and -1/3 are given for d log [ ]/d log $p(O_2)$, with p, $[V_M'']$, and $[V_O^{\cdot\cdot}]$, respectively.

At low $p(O_2)$, the expressions (78)-(80) give

$$2[V_M''] = 2[V_O^{\cdot\cdot}] \quad (= 2\sqrt{K_S}) \tag{81}$$

and

$$\frac{d \log p}{d \log p(O_2)^{1/2}} = \frac{1}{2} \tag{82}$$

It can be seen that Eq. (81) is an alternative approximation for the neutrality condition, Eq. (71).

In summary, at high $p(O_2)$ the fuller expression for electroneutrality, Eq. (77), approaches one approximation [Eq. (72)], and at low $p(O_2)$ it approaches another [Eq. (81)]; i.e., the condition $2[V_M''] = p$ changes to $2[V_M''] = 2[V_O^{\cdot\cdot}]$ as the oxygen pressure is reduced: the two species having effective charges of the same sign in the fuller expression [Eq. (77)] are interchanged as the neutrality approximation appropriate to one range of $p(O_2)$ changes into that appropriate to the next. It is seen that each simple two-term approximation, such as Eqs. (72) and (81), defines a new "range" or span of oxygen pressures in which the relations between the defect concentrations and the oxygen pressure take on a new form.

Since the mass-action equations are always valid, they apply equally well in all the ranges; the distinction between one range and the next is that a different approximation to the neutrality condition is used in the two cases. When the approximation is identified the concentrations of the defects and their dependence on $p(O_2)$ are readily found, as noted above.

Consider the new, low-p(O_2) approximation, Eq. (81); this is Schottky disorder, and in this range the vacancy concentrations are equal and independent of oxygen pressure. The dependence of p and n on p(O_2) in this range may be found from Eqs. (69), (70), and (81): $[V_M'']p^2 = K_A\, p(O_2)^{1/2}$, so

$$p = \left(\frac{K_A}{\sqrt{K_S}}\right)^{1/2} p(O_2)^{1/4} \qquad (83)$$

and

$$n = \frac{K_i}{p} = K_i\left(\frac{K_A}{\sqrt{K_S}}\right)^{-1/2} p(O_2)^{-1/4} \qquad (84)$$

The concentrations are shown in the central part of Fig. 9(a). As p(O_2) is lowered still further, the concentration n of electrons rises and, by analogy with the above, we pass from the range where $[V_M''] = [V_O^{\cdot\cdot}]$, through an intermediate zone where the values of n and $[V_M'']$ are similar, into a new range where the approximation

$$n = 2[V_O^{\cdot\cdot}] \qquad (85)$$

applies. Using Eq. (85), the rest of Fig. 9(a) (left-hand side) can be completed for this new range.

In the case of Fig. 9(b), for the right-hand range, $n > [V_O^{\cdot\cdot}]$. The neutrality approximation for the central range is then n = p, which with Eq. (69) gives $n = p = \sqrt{K_i}$ for that range. The same type of argument as used above will complete the diagram as shown.

The example considered shows two important features of defect equilibria. In the first place, defect behavior can be divided into ranges, which are spans of oxygen pressure within which a particular two-term approximation to the neutrality condition applies. The ranges are separated by zones in which a neutrality expression with three terms applies; however, in drawing Fig. 9, it has been assumed that one passes directly from one range into the next. The error introduced by this assumption is relatively small. From Eqs. (78)-(80) it can be seen that the defect concentrations pass smoothly from one range to the next, rather than with the abrupt transitions of Fig. 9; also a maximum error of a factor of two occurs in the concentrations at the transition points. Examples of concentration diagrams drawn without the above assumption are given in the paper by Kröger and Vink [4].

The second important feature to be noted in the diagram is that the behavior of the material and with it the form of diagram is dependent on the values of the equilibrium constants for the various reactions involving the defects; a comparison of parts (a) and (b) of Fig. 9 emphasizes this fact.

The approach used for the pure compound MO is readily extended to cover the effects of impurities. Let us add a species N which enters the lattice substitutionally on M sites and produces levels N_M^x and $N_M^{\bullet}$ in the conduction and valence bands, respectively, so that it mainly exists in the crystal as $N_M^{\bullet}$. Then we have an added unknown $[N_M^{\bullet}]$. Against this we have one of two situations.

(a) There may be an incorporation mechanism for N in terms of the defects present, e.g.,

$$N_{(g)} \rightarrow N_M^{\bullet} + V_O^{\bullet\bullet} + 3e' \tag{86}$$

This would be used in cases where the impurity is present in the crystal in equilibrium with the impurity species present in the gas phase.

(b) Without an incorporation mechanism, there must be a mass-balance expression for N, namely,

$$[N]_{Total} = \text{const.}$$
$$= \Sigma[\text{defect types containing N}] \tag{87}$$

For example, one might under other circumstances have N present as $N_M^{\bullet}$ and $N_i^{\bullet\bullet\bullet}$; then

$$[N]_{Total} = [N_M^{\bullet}] + [N_i^{\bullet\bullet\bullet}] \tag{88}$$

Alternative (b) would apply where a fixed impurity concentration occurs within the crystal under quasi-equilibrium conditions; the impurity does not exchange with the environment, but it does fully participate in the defect equilibria within the crystal.

The presence of the additional unknown $[N_M^{\bullet}]$ is therefore accompanied by an additional equation involving the defects, and the procedure adopted for defects in the pure oxide can apply also for the impure system. The incorporation reaction for N [Eq. (86)] gives a mass-action equation which can be expressed as a linear relation between the logarithms of the defect concentrations. The mass-balance equation (88) is treated in the same way as the neutrality condition (or charge-balance equation), being approximated by the dominant term on either side; the resulting two-term equation can then be written as a linear relation between the logarithms of the concentrations of the dominant terms. The Brouwer diagram can then be constructed as outlined above for the pure system. Since $N_M^{\bullet}$ carries an effective charge, it must be added to the terms on the right-hand side of the full neutrality condition [Eq. (71)].

The effect of having $[N_M^{\bullet}]^2 > K_S$, K_i is shown in Fig. 10 for the cases corresponding in the pure state to Fig. 9; the neutrality approximations are shown for each range.

The remaining effect to consider in diagrams of this type is that of defects having electron levels within the gap. As an example, consider the case of the level diagram of Fig. 11(a). Since the defect levels $V_O^{\bullet}$ and V_M'' are now within the gap, changes in the Fermi level [brought about by changes in $p(O_2)$] can empty or fill these levels. In addition to $[V_O^{\bullet\bullet}]$ and $[V_M'']$, therefore, we have $[V_O^{\bullet}]$ and $[V_M']$ as two additional unknowns. Against these unknowns, for each level occurring within the gap, we have an ionization reaction and a corresponding mass-action expression; for the present example these are

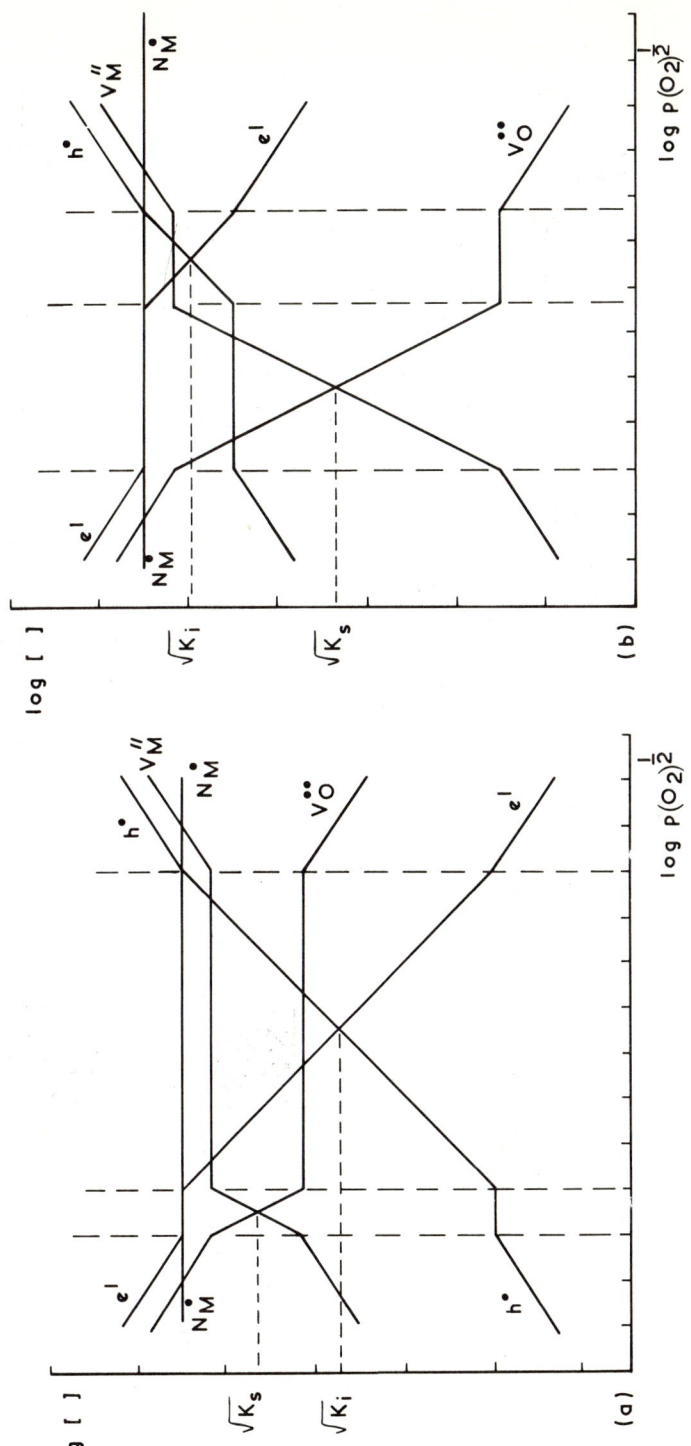

FIG. 10. The effect of added impurity $N_M^\cdot$ on the defect situations of Fig. 9. Note that horizontal scale = $\frac{1}{2}$ vertical scale. Neutrality approximation ranges from left to right are $n = 2[V_O^{\cdot\cdot}]$, $n = [N_M^\cdot]$, $[N_M^\cdot] = 2[V_M'']$, and $p = 2[V_M'']$. (a) $K_S > K_i$ [Fig. 9(a)]. (b) $K_S < K_i$ [Fig. 9(b)].

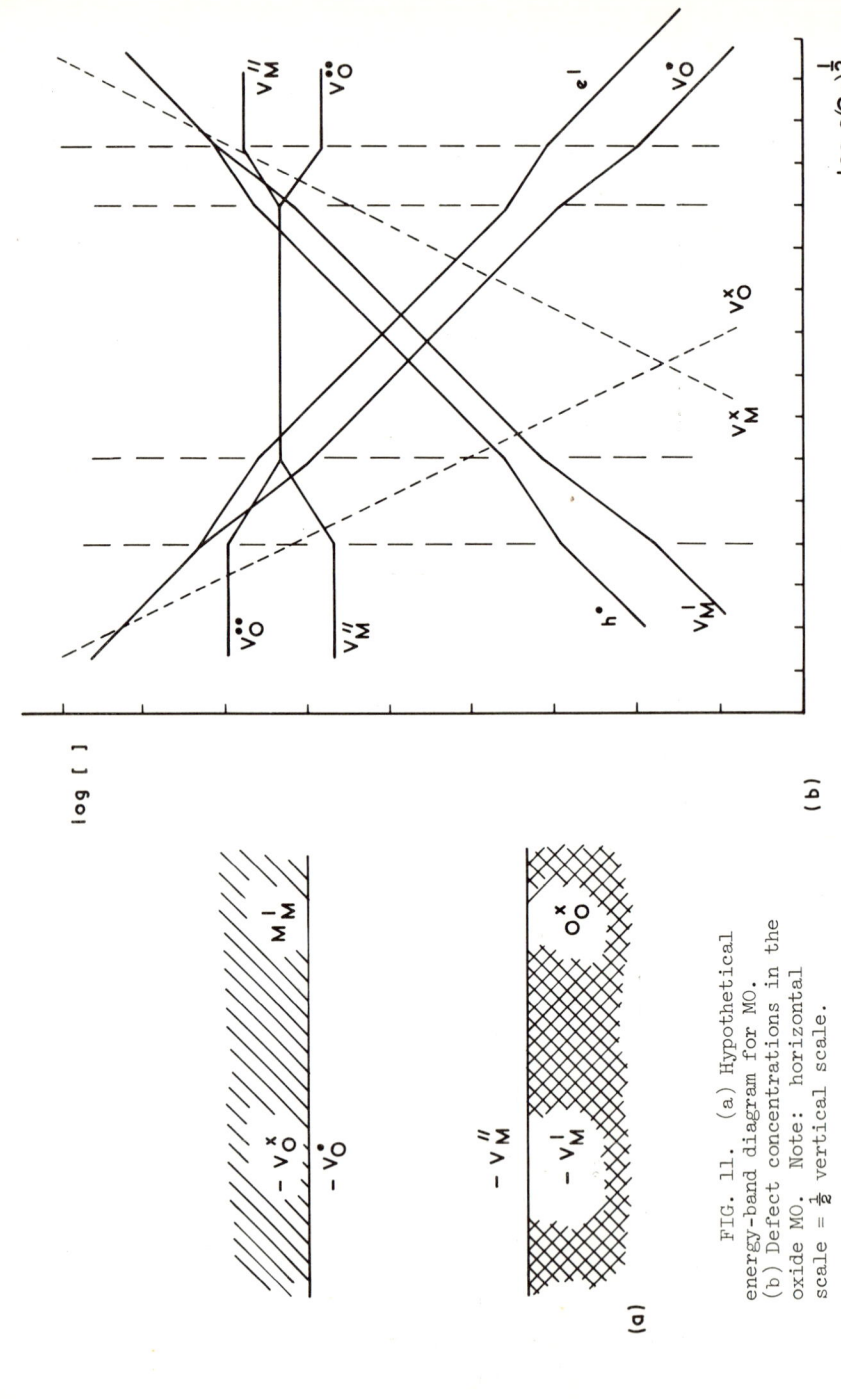

FIG. 11. (a) Hypothetical energy-band diagram for MO. (b) Defect concentrations in the oxide MO. Note: horizontal scale = ½ vertical scale.

DEFECT STRUCTURE OF CERAMIC MATERIALS

$$V_O^{\cdot} \rightarrow e' + V_O^{\cdot\cdot}$$

and

$$V_M'' \rightarrow V_M' + e' \quad (\text{or } V_M' \rightarrow h^{\cdot} + V_M'') \tag{89}$$

The concentrations $[V_O^{\cdot}]$ and $[V_M']$ must also be added to the full neutrality condition. The Brouwer diagram for this situation is shown in Fig. 11(b).

The ionization of an impurity (resulting from the occurrence of an impurity level with the gap) is treated in an identical manner, the new concentration value being offset by the new ionization equation. The method is readily extended to more complex situations, resulting in diagrams with more ranges: the form of the diagram basically depends as noted on the values of the constants for the mass-action equations.

The results of bringing the V_M' and V_O^x levels into the gap (see Fig. 6, right-hand-side levels) is to allow two new defect types, V_M^x and V_O^x; these are neutral and hence do not appear in the neutrality condition or influence the form of the Brouwer diagram for the other defect species. Their concentrations can be found by way of two additional ionization reactions (e.g., $V_M^x \rightarrow h^{\cdot} + V_M'$) or by way of the direct incorporation reactions (33) and (36)

$$\tfrac{1}{2}O_{2(g)} \rightarrow V_M^x + O_O^x$$

$$\tfrac{1}{2}O_{2(g)} + V_O^x \rightarrow O_O^x$$

whence we may derive Eqs. (35) and (38):

$$[V_M^x] = K_2 p(O_2)^{1/2}$$

$$[V_O^x] = K_3 p(O_2)^{1/2}$$

These are shown as dotted lines on Fig. 11(b).

2. Effects of Impurity Concentration Variation

In the preceding discussion we have considered how the different defect concentrations have varied with the changing composition of the crystal and with variations in the oxygen pressure. The effect of changing impurity concentration can be evaluated in a similar manner. For instance, the effect of adding $N_M^{\bullet}$ to the crystal of Fig. 9(a) is shown in Fig. 12(a). At low $N_M^{\bullet}$, the full neutrality condition

$$[N_M^{\bullet}] + p + 2[V_O^{\bullet\bullet}] = 2[V_M''] + n \qquad (90)$$

is, for the oxygen pressure chosen, approximated by Eq. (81)

$$2[V_O^{\bullet\bullet}] = 2[V_M'']$$

As $N_M^{\bullet}$ rises, a new range appears for which the approximation is

$$[N_M^{\bullet}] = 2[V_M''] \qquad (91)$$

The concentrations of the other species are then calculated from Eqs. (68), (69), and (70). Figure 12(b) corresponds to adding $N_M^{\bullet}$ to the crystal of Fig. 9(b), at an oxygen pressure where the neutrality approximation in the pure crystal is $n = p$.

3. Pressure Dependence of Individual Defects

While the Brouwer diagrams present a complete picture of the defect behavior of the crystal, it is often convenient to evaluate the defect behavior within a single range, since measurements frequently cover a restricted span of oxygen pressures. In such cases drawing the full diagram is unnecessary and direct use of the mass-action laws may be made. To calculate the pressure dependence of a certain defect concentration, the following scheme may be used.

(1) Identify the two species appearing in the approximation to the neutrality condition.

(2) Evaluate the dependence of these two species on the oxygen pressure. (a) For intrinsic disorder and for a neutrality approximation involving a fixed concentration of impurity, the species concentrations are independent of oxygen pressure. (b) For extrinsic disorder, the dependence can be calculated by writing mass-

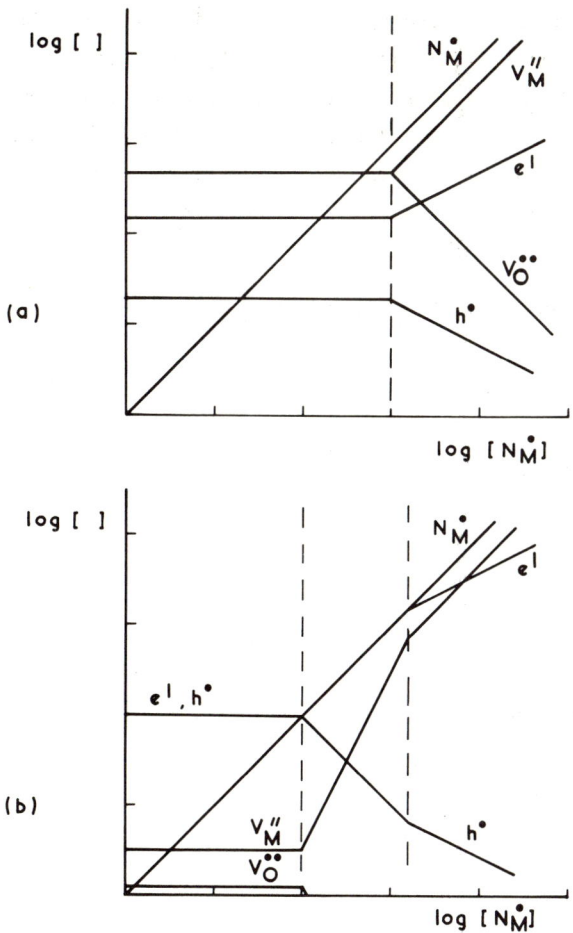

FIG. 12. (a) The effect of adding $N_M^\bullet$ to the defect situation of Fig. 9(a); $p(O_2)$ = const. (b) The effect of adding $N_M^\bullet$ to the defect situation of Fig. 9(b); $p(O_2)$ = const. Note that the first neutrality range involving the impurity includes also a species from the neutrality approximation of the pure crystal; deliberately added dopants can in this way be used to reveal part of the defect structure in the pure crystal. Where multi-ply charged defects are involved, it is important not to overlook the possible existence of another impurity-dominated region [see part (b)].

action equations for the inclusion reaction in terms of the oxygen pressure, the concentrations of both the species in the neutrality approximation and (where applicable) a fixed neutral impurity concentration (or a fixed external impurity pressure).

(3) The dependence of any other defect species on the oxygen pressure can then be found by writing an equation in terms of the concentration of the species of interest, the concentration of one of the species in the neutrality approximation and (if necessary) the oxygen pressure.

For example, in the right-hand range of Fig. 11(b), we have

$$p = [V'_M]$$

Then since

$$\tfrac{1}{2}O_2(g) \rightarrow O^x_O + V'_M + h^\cdot$$

$$p[V'_M] = K p(O_2)^{1/2}$$

and

$$p = [V'_M] = K^{1/2} p(O_2)^{1/4} \qquad (92)$$

The dependence of $[V''_M]$, for example, is then given by

$$\tfrac{1}{2}O_2(g) \rightarrow V''_M + O^x_O + 2h^\cdot$$

whence $p^2[V''_M] = K_A p(O_2)^{1/2}$ [Eq. (70)].

From Eqs. (92) and (70),

$$[V''_M] = \text{const.} \qquad (93)$$

4. Effect of Temperature Variation

If we assume that the standard entropy changes and standard enthalpy changes of the various defect reactions are independent of temperature, then from Eq. (21) we have that

$$K = \text{const.} \exp\left(-\frac{\Delta H^o}{RT}\right) \qquad (94)$$

The effect of temperature variation on the defect concentrations can then be calculated by adding this temperature dependence of the

DEFECT STRUCTURE OF CERAMIC MATERIALS   229

equilibrium constant to the types of calculation performed in Sections V,A,1-2 above.

As an example, consider the situation of Fig. 9(b), and let the oxygen pressure be constant and such that the system is in the central range, where n = p. Here Eqs. (68)-(71) still apply, but the equilibrium constants are no longer constant but given by

$$K_S = K_S^0 \exp(-\frac{\Delta H_S^0}{RT}) \tag{95}$$

$$K_i = K_i^0 \exp(-\frac{\Delta E_i}{RT}) \tag{96}$$

$$K_A = K_A^0 \exp(-\frac{\Delta H_A^0}{RT}) \tag{97}$$

Since n = p, from Eq. (69) we have

$$n = p = \sqrt{K_i} = \sqrt{K_i^0} \exp(-\frac{\Delta E_i}{2RT})$$

and a plot of $\log_e n$ (or $\log_e p$) vs $1/T$ will give a straight line of slope $-\Delta E_i/2R$. Once the temperature dependence of the species in the neutrality approximation is known, the dependence of the others can be determined. For instance, from Eq. (70)

$$[V_M''] = \frac{K_A p(O_2)^{1/2}}{p^2} \tag{98}$$

$$\simeq \frac{K_A^0 \exp(-\Delta H_A^0/RT)}{K_i^0 \exp(-\Delta E_i/RT)} p(O_2)^{1/2} \tag{99}$$

whence

$$\frac{d \log_e [V_M'']}{d(1/T)} = -\frac{\Delta H_A^0}{R} + \frac{\Delta E_i}{R} \tag{100}$$

Since the temperature dependence of the various defects can be different, the form of the neutrality approximation can change as a function of temperature, leading to a number of ranges much as was found for oxygen-pressure variations. The full diagram for the present case is shown in Fig. 13.

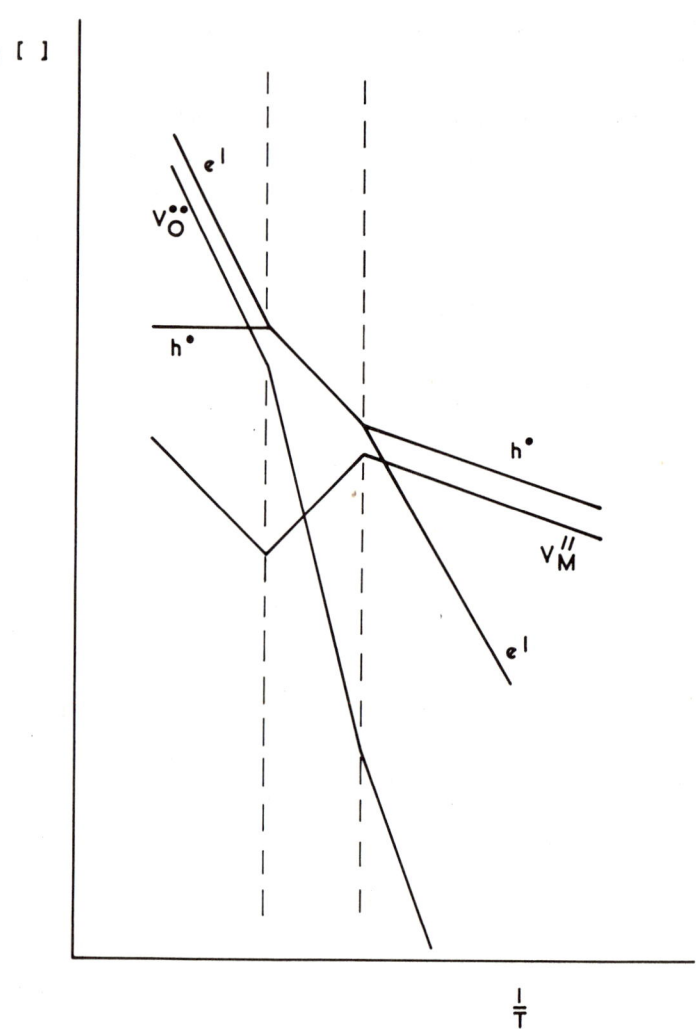

FIG. 13. The effect of temperature on the defect situation of Fig. 9(b); $p(O_2)$ = const. $\Delta H_A^O : \Delta H_S^O : \Delta E_i = 1 : 3 : 2$.

An important example of temperature-dependent defect behavior can occur when impurity defects are involved in the neutrality approximation, e.g., as in Fig. 12(a), right-hand range. If the impurity is present as a fixed "quenched-in" concentration, rather

DEFECT STRUCTURE OF CERAMIC MATERIALS

than in equilibrium with some species of N in the surrounding gas phase, $[N_M^\cdot]$ is constant and independent of the temperature. Since, in the example,

$$[N_M^\cdot] = 2[V_M''] = \text{const.} \qquad (101)$$

we have from Eq. (68)

$$[V_O^{\cdot\cdot}] = \frac{K_S}{[V_M'']}$$

$$= \frac{K_S^o \exp(-\Delta H_S^o/RT)}{\frac{1}{2}[N_M^\cdot]} \qquad (102)$$

and

$$\frac{d \log_e [V_O^{\cdot\cdot}]}{d(1/T)} = -\frac{\Delta H_S^o}{R} \qquad (103)$$

As the temperature is raised, $[V_O^{\cdot\cdot}]$ rises and eventually a new neutrality approximation, $2[V_M''] = 2[V_O^{\cdot\cdot}]$, becomes valid. This is shown in Fig. 14. For the new range the temperature dependence is given by Eqs. (68) and (95)

$$[V_M''] = [V_O^{\cdot\cdot}] = \sqrt{K_S^o} \exp\left(-\frac{\Delta H_S^o}{2RT}\right) \qquad (104)$$

The diagram is completed as for the other example. It may be noted that as the temperature is raised the system passes from an impurity-controlled extrinsic behavior to a region of intrinsic behavior.

5. Quenched Systems

With the exception of the quenched-in impurity concentrations discussed earlier, this treatment of defect equilibria in crystals has been concerned with defects in thermal equilibrium. In some cases it is of interest to know how the equilibria established at high temperature affect measurements made on materials at lower temperature, where equilibrium no longer pertains.

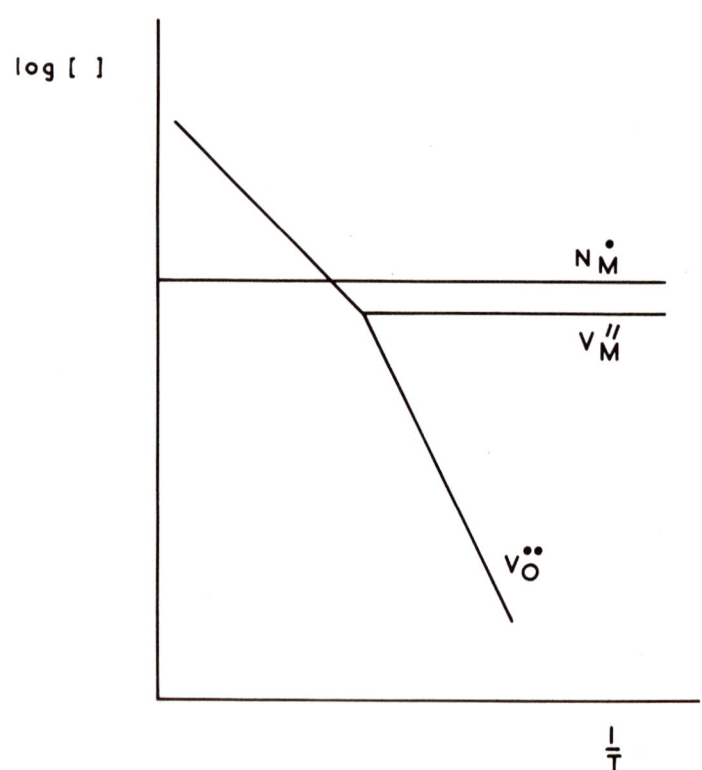

FIG. 14. The effect of temperature on a defect situation involving impurities; $p(O_2)$ = const.

An extensive treatment can be found in the book by Kröger [3]; here it can be noted that since the time constants for equilibration of electronic and atomic defect concentrations are short and long, respectively, a common approximation is that the atomic-defect concentrations remain, when quenched, at their high-temperature level, whereas the electronic defects continue to equilibrate. Then, as before, mass-action equations are written for the equilibrated species, and balance equations, analogous to Eqs. (40) and (88), are written for the quenched-in defects; the methods of treating the two types of equation are summarized in the discussion following Eq. (88).

6. Multicomponent Systems

For a binary compound, e.g., MO in the preceding discussion, the control of one component [$p(O_2)$] in the surrounding atmosphere is sufficient to define the composition of the solid. For multicomponent systems, use of the phase rule demonstrates that each additional component requires control of an additional compositional variable if the system is to be completely defined.

Many important ceramic materials (spinels and perovskites) are at least three-component systems; possible compositional factors that can be controlled are, for instance, the oxygen pressure and the compositional ratio of the cations. For example in AB_2O_4, this means control of $p(O_2)$ and A/B. The system is then completely defined.

For an extensive treatment of this problem, see the paper by Schmalzried [13].

B. LARGE DEFECT CONCENTRATIONS

In Section V,A, the defect concentrations have been assumed to be small; this has meant

(a) that the mass-action equations may be written in terms of concentrations rather than activities (interaction between defects can be neglected), and

(b) that the concentrations of the normal lattice constituents such as $[V_i^x]$, $[M_M^x]$, and $[O_O^x]$ may be considered constant, and unaltered by the creation of defects.

When larger concentrations of defects are encountered these approximations are no longer valid, and a number of refinements can be applied. These may be considered in turn. Comparison of the various treatments with experimental results will show which, if any, of the refinements is suitable for the given system and experimental conditions.

1. Variation in the Concentration of Normal Lattice Constituents

As a first step, we may still assume that interactions between defects are negligible, so that concentrations may still be used in the mass-action expression, but we take account of the effect that the introduction of defects has on the concentration of the lattice constituents.

As an example, consider MO to be a transition-metal oxide, so that, as for NiO in Fig. 5, the M^+ band is the valence band; if the band is narrow, electron holes are localized and can be formally represented by M^{3+} ions ($M_M^{\cdot}$) in the solid. A reaction for incorporation of oxygen then reads

$$2M_M^x + \tfrac{1}{2}O_2 \rightarrow O_O^x + V_M'' + 2M_M^{\cdot} \tag{105}$$

In the mass-action treatments above, we have assumed that the site fractions $[O_O^x]$ and $[M_M^x]$ of the lattice constituents were negligibly different from unity considering the small concentrations of defects involved; hence Eq. (105) led to

$$[V_M''][M_M^{\cdot}]^2 = K_A p(O_2)^{1/2} \tag{106}$$

If large concentrations of defects are formed by reaction (105), however, $[M_M^x]$ may be considerably below unity, since for each O_O^x added two M_M^x are converted to $M_M^{\cdot}$, and the value of $[M_M^x]$, given by

$$\frac{\varphi(M_M^x)}{\varphi(M_M^x) + \varphi(M_M^{\cdot}) + \varphi(V_M'')}$$

where $\varphi()$ denotes "number of", falls. For reaction (105), note that $[O_O^x]$ is unchanged. Consequently, a more accurate mass-action expression is

$$[V_M''][M_M^{\cdot}]^2 = K_A p(O_2)^{1/2}[M_M^x]^2 \tag{107}$$

where $[M_M^x]$ is no longer considered as constant. The oxygen pressure dependences of the defect concentrations given by Eqs. (106) and (107) are different.

DEFECT STRUCTURE OF CERAMIC MATERIALS   235

For example, for the neutrality approximation $2[V_M''] = [M_M^\bullet]$, Eq. (106) gives

$$[V_M''] = (\frac{K_A}{4})^{1/3} p(O_2)^{1/6} \qquad (108)$$

so that

$$\frac{d \log p(O_2)^{1/2}}{d \log [V_M'']} = 3 \qquad (109)$$

In contrast, Eq. (107) gives

$$[V_M'']^3 = \frac{K_A}{4} p(O_2)^{1/2} [M_M^x]^2 \qquad (110)$$

Since, using site fractions,

$$[M_M^x] = 1 - [M_M^\bullet] - [V_M'']$$

$$= 1 - 3[V_M''] \qquad (111)$$

we have

$$p(O_2)^{1/2} = \frac{(4/K_A)[V_M'']^3}{(1 - 3[V_M''])^2} \qquad (112)$$

and

$$\frac{d \log p(O_2)^{1/2}}{d \log [V_M'']} = 3 + \frac{6[V_M'']}{1 - 3[V_M'']} \qquad (113)$$

Equation (109) is seen to be an approximation for Eq. (113), valid at small defect concentrations; it is 10% in error when $[V_M''] \simeq 0.06$, a large, but not unknown (see Table 2), defect concentration.

According to Eq. (113), the slope of the defect concentration line on the Brouwer diagram is dependent on the value of the concentration; consequently, the use of the diagrammatic representation of defect situations becomes impractical, and the discussion is generally restricted to defect behavior within a single range.

In a similar manner, in nonstoichiometry involving the formation of interstitial defects, the vacant interstitial sites can be introduced as lattice constituents of varying concentration. For example, in the case of Eq. (32),

$$\tfrac{1}{2}O_{2(g)} + V_i^x \rightarrow O_i^x$$

we should, at large deviations from stoichiometry, use

$$[V_i^x] = 1 - [O_i^x] \tag{114}$$

and not assume, as in deriving Eq. (34), that

$$[V_i^x] = 1 \tag{115}$$

2. Interactions Among Defects

When the defect concentration reaches the level where interactions between the defects become important, we no longer have an ideal solution of defects in the crystal; there are two techniques whereby we can still attempt to use concentrations in the mass-action equations.

a. *Associates*. When the defects interact over a short range, either as a result of coulombic or chemical forces, the situation can be represented by treating the closely interacting defects as a new species, termed an associate. This may then be used in the mass-action equations as a new defect, the equations being written, as before, in terms of concentrations. Examples of this method have been seen earlier in the discussion of ionization of defects; for instance, reaction (55)

$$M_i^x \rightarrow e' + M_i^\bullet$$

(the ionization of M_i^x) may be considered alternatively as representing the interaction of two defects to form an associate which is then in turn treated as a defect in the mass-action equations:

$$e' + M_i^\bullet \rightarrow M_i^x \; ; \quad \frac{[M_i^x]}{n\,[M_i^\bullet]} = K_{M_i^x}^{-1} \tag{116}$$

DEFECT STRUCTURE OF CERAMIC MATERIALS

In this example, there is a coulombic interaction between the two component defects, e' and $M_i^{\cdot}$. Chemical or mechanical forces can cause association between defects where no such coulombic interaction is involved.

A more complex example of a defect associate arises from combination of the products of reaction (27)

$$M_M^x + V_i^x \to M_i^x + V_M^x$$

and reaction (33)

$$\tfrac{1}{2}O_{2(g)} \to O_O^x + V_M^x$$

to give

$$M_i^x + V_M^x + V_M^x \to (V_M M_i V_M)^x \tag{117}$$

Since the heat of association for attracting defects is negative, association becomes less pronounced as the temperature is raised. At low temperatures, the free defect concentrations may well be negligible, so that the incorporation reaction for oxygen in the above example may become

$$M_M^x + V_i^x + \tfrac{1}{2}O_{2(g)} \to O_O^x + (V_M M_i V_M)^x$$

$$[(V_M M_i V_M)^x] = Kp(O_2)^{1/2} \tag{118}$$

A similar type of defect has been suggested to account for the properties of some of the transition-metal oxides; where associates consist of a number of primary types, as in this case, they are often termed "complex defects".

A further example of association is shown in Fig. 15.

b. <u>Steric or Repulsive Effects</u>. The occurrence of short-range repulsive forces may also be accounted for in the mass-action equations. Let us again take as example the incorporation reaction (32)

$$\tfrac{1}{2}O_{2(g)} + V_i^x \to O_i^x$$

but let there now be repulsive forces between O_i^x defects such that

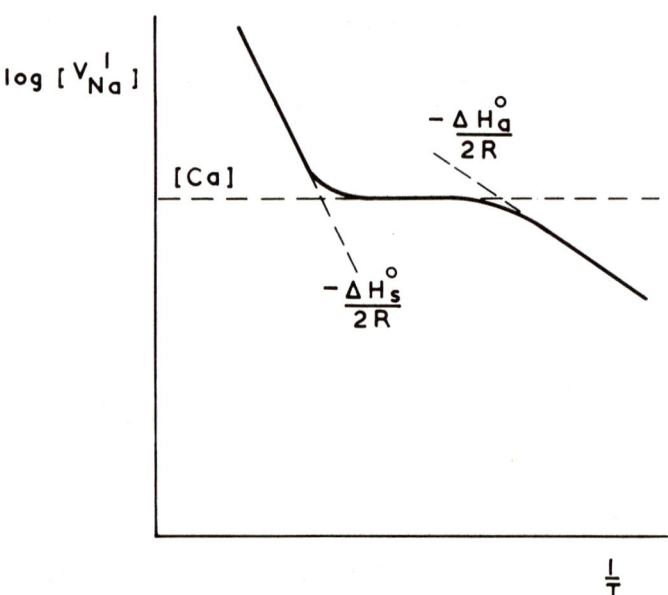

FIG. 15. The temperature dependence of the concentration of sodium vacancies in calcium-doped sodium chloride. The three linear regions represent, from left to right, $[V'_{Na}] = [V^{\cdot}_{Cl}]$; $[V'_{Na}] = [Ca^{\cdot}_{Na}]$ = const.; $[V'_{Na}] = [Ca^{\cdot}_{Na}] = f(T)$. In the right-hand region the temperature-dependent concentration of $Ca^{\cdot}_{Na}$ arises from the presence of an associate

$$(Ca^{\cdot}_{Na} V'_{Na})^x \rightarrow Ca^{\cdot}_{Na} + V'_{Na} \; ; \; \Delta H_a$$

$$\frac{[Ca^{\cdot}_{Na}][V'_{Na}]}{[(Ca^{\cdot}_{Na} V'_{Na})]} \simeq \exp\left(-\frac{\Delta H^o_a}{RT}\right)$$

they cannot occupy neighboring interstitial sites. Then in place of Eq. (114) we should use

$$[V^x_i] = 1 - [O^x_i] - Z[O^x_i] \tag{119}$$

where Z is the number of interstitial sites surrounding another interstitial site. In other words, each O^x_i occupies one V^x_i site, and makes Z V^x_i sites unsuitable for occupation by other interstitials.

Besides allowing for repulsive forces between defects, this approach allows treatment of steric effects; e.g., where a complex defect is located at a site within the crystal, it may prevent access of other complex defects to neighboring sites on geometrical grounds only. An example is contained in the paper by Libowitz [14].

3. The Use of Activity Terms

When the forces between defects are long-range coulombic interactions, and when the departures from a random distribution of the defects in the solid are also long-range, each effectively positive species has a cloud of predominantly negative effective charges around it. The departures from ideality must then be accounted for by introducing activities in place of concentrations in the mass-action equations; the activities can for small concentrations be derived from the Debye-Hückel theory, which was originally developed to account for similar interactions in ionic liquids.

As an example, we can consider the reaction (116)

$$e' + M_i^{\cdot} \to M_i^x$$

Since the initial stage of the association reaction between the two oppositely charged defects (prior to their forming the associate) is long-range, the mass-action equation corresponding to reaction (116) should strictly use Debye-Hückel corrections for the charged defects, resulting in the expression

$$(K_{M_i^x}) = \frac{[M_i^x]}{f_e^n \; f_{M_i^{\cdot}}[M_i^{\cdot}]}$$

for the equilibrium constant; f_e and $f_{M_i^{\cdot}}$ are activity coefficients calculated using the Debye-Hückel theory.

A discussion of this topic can be found in the paper by Lidiard [15].

C. EXTENDED DEFECTS

1. Microdomains and Precipitates

We have described above how one result of the defect interactions that stem from large defect concentrations can be the occurrence of associates. In some cases these associates are quite complex assemblies of the primary defects, such as the $(V_M M_i V_M)^x$ of the example. As the concentration of such complex defects is raised, there will in turn be interactions among these, leading to further association and eventually to regions in the crystal having a structure related to that of the defect complex rather than to that of the original crystal. These regions have been termed microdomains and can be considered early stages in the phase transformation from the original crystal structure to that represented by the associated complexes.

Crystals which are strongly nonstoichiometric can in this picture be considered as systems with extensive development of the microdomains; the same process in crystals which are strongly impure is represented by association of the impurities, and leads to precipitation of a second impurity phase.

2. Shear Structures

An alternative method for incorporating large degrees of nonstoichiometry or impurity occurs in some crystals, and involves the formation of shear structures. We have noted that an ordering of point defects can occur as a result of defect interactions, either by association of complexes or by a combination of large point-defect concentrations with repulsive effects of the type described in Section V,B,2,b above. When the point defects are ordered onto a superlattice in this way, the crystal can, if it is of the appropriate type, eliminate a plane of the ordered defects by undergoing crystallographic shear (Fig. 16). By further operation of this mechanism, the crystal becomes nonstoichiometric by creating planar faults rather than by creating point defects.

DEFECT STRUCTURE OF CERAMIC MATERIALS

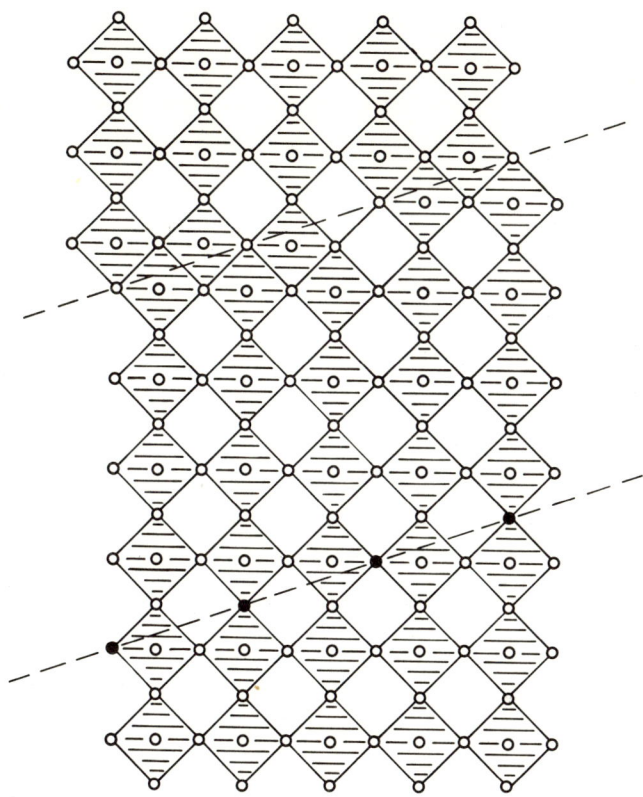

FIG. 16. The Anderson and Hyde model for the production of a $(1\bar{3}0)$ crystallographic shear plane in ReO_3. Oxygen vacancies aggregate into a planar disk across which shear occurs to produce an element of a crystallographic shear plane bounded by a partial dislocation loop. The loop extends by dislocation climb as the vacancies accrete. [Reprinted from Ref. 6 by courtesy of North-Holland Publishing Company.]

At high concentrations of planar faults (Wadsley defects), they can become ordered, and the result is a shear structure; this is a crystal similar to the original stoichiometric form, but incorporating regularly spaced planes across which crystallographic shear has occurred.

The two forms of extended defect noted here fall legitimately in the area of defects in ceramics; however, their influence on ion movements and on electrical conductivity, though not yet well investigated or understood, is likely to be less significant than that of the point defects. Their main importance lies in their ability to introduce substantial degrees of nonstoichiometry into the system in which they occur.

Further discussion and many examples of shear-structure formation can be found in the book by Eyring and O'Keeffe [6].

VI. MICROSTRUCTURAL EFFECTS

In all the earlier discussion we have been treating the properties of an infinite, dislocation-free, single crystal. Since we know that the system must be overall charge-neutral, this has meant that any effectively charged defects have to be compensated by oppositely charged defects, a state of affairs that is recognized in the neutrality condition.

In real ceramic systems there are usually surfaces, dislocations, and grain boundaries, and since these can themselves carry effective electrical charges, we can achieve an overall charge neutrality without having exactly compensating concentrations of point defects. Some methods in which dislocations and surfaces can carry charge are shown in Fig. 17.

If we consider as example an oxide MO, which, in the bulk and in the absence of line or plane defects, exhibits Schottky disorder, then we have the situation described by Eq. (81)

$$[V_M''] = [V_O^{\cdot\cdot}]$$

as the neutrality approximation, and

$$[V_M''][V_O^{\cdot\cdot}] = \exp\left(-\frac{\Delta G_S^o}{RT}\right) \tag{120}$$

as the relation for defect concentrations.

DEFECT STRUCTURE OF CERAMIC MATERIALS 243

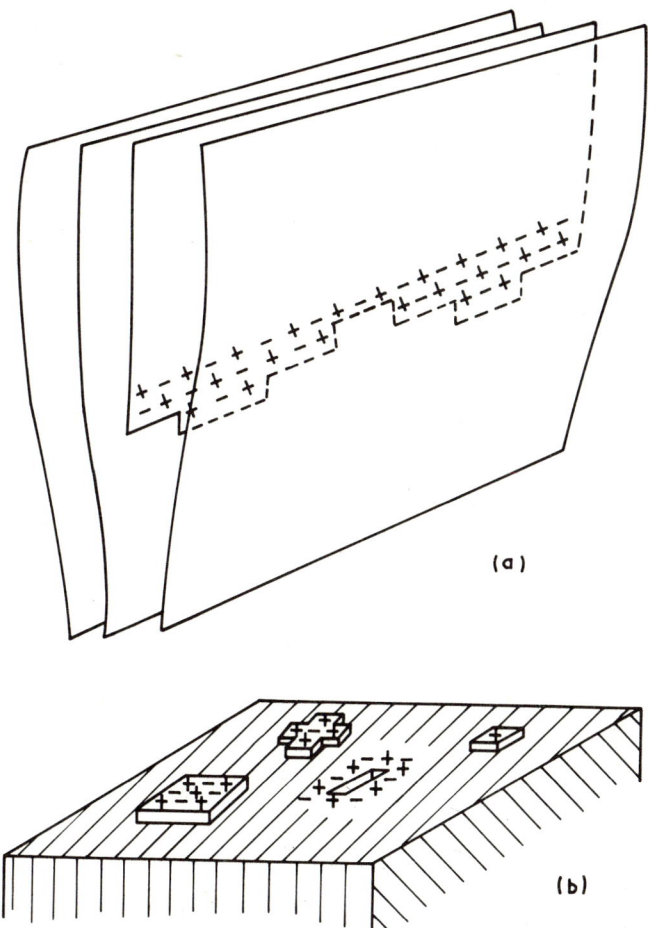

FIG. 17. Schematic pictures of (a) a positively charged edge dislocation, and (b) a positively charged crystal surface.

Consider an initially neutral dislocation located in the bulk of the crystal. If the energy to form V_M'' is less than the energy to form $V_O^{\cdot\cdot}$, then in the absence of the neutrality requirement there would be a tendency for $[V_M''] > [V_O^{\cdot\cdot}]$. The presence of the dislocation allows this inequality; M^{2+} ions can be absorbed onto jogs on the dislocation ($M_d^{\cdot\cdot}$) with the creation of excess V_M'' defects around the dislocation. At equilibrium the dislocation exists as a positively charged line defect surrounded by a negatively charged cloud of V_M''

defects, whose tendency to diffuse away from the dislocation down the concentration gradient is counteracted by the coulombic attraction toward the dislocation.

Since Eq. (120) still applies, the region around the dislocation will be relatively depleted in $V_O^{\cdot\cdot}$. The overall picture is shown in Fig. 18. Note that the region as a whole remains neutral according to the equation

$$2[V_O^{\cdot\cdot}] + 2[M_d^{\cdot\cdot}] = 2[V_M''] \qquad (121)$$

Local neutrality is of course not maintained, the charge distribution around the dislocation being given exactly as for the Debye-Hückel cloud around point defects.

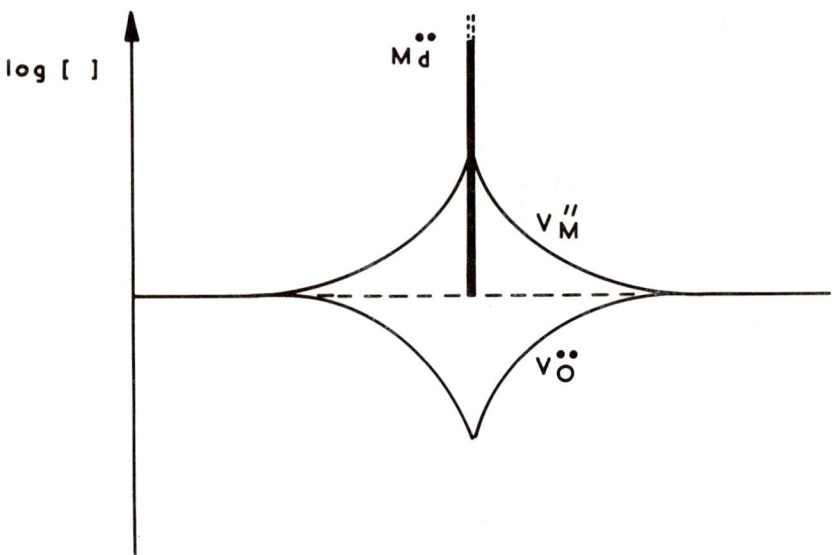

FIG. 18. The concentration of point defects around a positively charged dislocation.

In deciding which species are absorbed onto the dislocation we can make use of the concept of "naive" defect concentrations, which

DEFECT STRUCTURE OF CERAMIC MATERIALS

are those that would occur in the bulk if the neutrality condition were unimportant. They are

$$[V_M'']_n = \exp\left(-\frac{G_c}{RT}\right) \tag{122}$$

and

$$[V_O^{\cdot\cdot}]_n = \exp\left(-\frac{G_a}{RT}\right) \tag{123}$$

where G_c and G_a are the energy for the creation of a cation vacancy by absorption of a cation onto the dislocation, and the energy of the same process for an anion. Note that

$$\Delta G_s^0 = G_c + G_a \tag{124}$$

In the earlier arguments, we supposed $G_c < G_a$; consequently, since this means that

$$G_c < \frac{G_c + G_a}{2} < G_a \tag{125}$$

we have

$$[V_M'']_n > [V_M'']_{bulk} = [V_O^{\cdot\cdot}]_{bulk} > [V_O^{\cdot\cdot}]_n \tag{126}$$

The presence of the dislocation allows a departure from the neutrality approximation (81), and a relaxation toward the naive concentrations. From Eq. (126), this leads to the occurrence of the situation shown in Fig. 18.

As a further development, consider that the crystal contains an impurity $N_M^{\cdot}$; then at low temperature (Fig. 14), the neutrality approximation becomes

$$[N_M^{\cdot}] = 2[V_M'']_{bulk} \tag{127}$$

On lowering the temperature, $[V_M'']_n$, $[V_O^{\cdot\cdot}]_n$, and $[V_O^{\cdot\cdot}]_{bulk}$ are reduced according to Eqs. (122), (123), and

$$[V_O^{\cdot\cdot}]_{bulk} = \frac{K_S}{[V_M'']} = \frac{2}{[N_M^{\cdot}]} \exp\left(-\frac{G_a + G_c}{RT}\right) \tag{128}$$

Since $(G_a + G_c) > G_a > G_c$, $[V_O^{\cdot\cdot}]_{bulk}$ falls more rapidly than the two naive concentrations. The end result is (see Fig. 19) that at low temperatures

$$[V_M'']_n < [V_M'']_{bulk} = \tfrac{1}{2}[N_M^{\cdot}] \tag{129}$$

and

$$[V_O^{\cdot\cdot}]_n > [V_O^{\cdot\cdot}]_{bulk} \tag{130}$$

so that the charge on the dislocation becomes negative. Figure 19 summarizes these features.

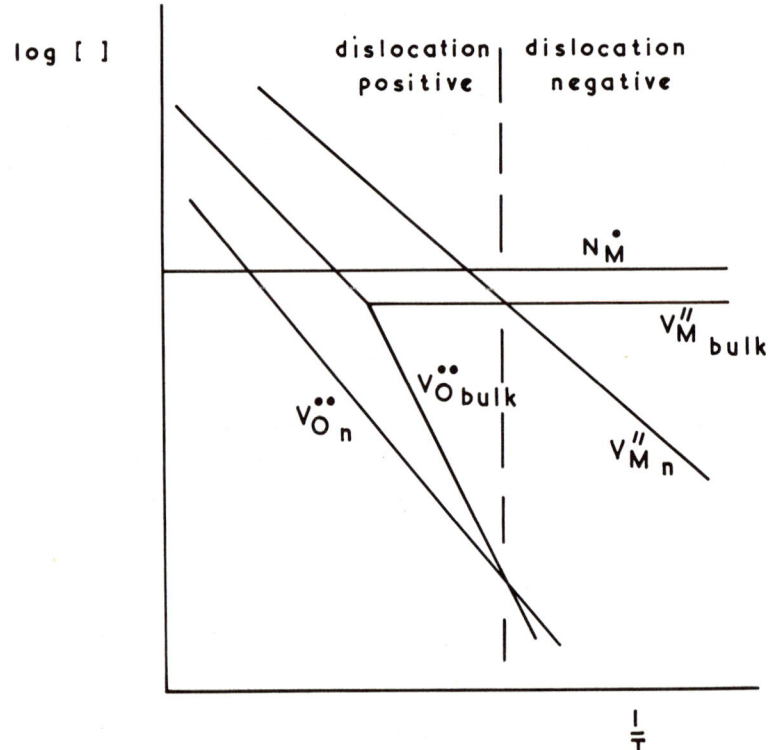

FIG. 19. The sign of the charge on dislocations in a crystal containing impurity, shown as a function of temperature.

The extent of the space-charge region is of the order of the Debye length $1/b$, where

$$b^2 = \frac{4\pi q^2}{\epsilon kT} \Sigma n_i z_i^2 \tag{131}$$

$$= \frac{32\pi q^2}{\epsilon kT} [V_M'']_{bulk} \tag{132}$$

Here ϵ is the dielectric constant, q the magnitude of the electronic charge, and z_i the number of units of charge carried by defect i. With reasonable values 1/b is about 1 µ.

While the above discussion has been in terms of dislocations, the same effects are found in the neighborhood of grain boundaries and of the outer surfaces of the crystal. Their importance is that these regions of enhanced defect concentrations are potentially regions of high defect (and hence ion) diffusion. What is not yet certain is whether the high-diffusion channel is the space-charge region or the region of the line (or plane) defect itself. With outer surfaces, it is probably the latter.

It should be emphasized that the above discussion has been concerned with equilibrium effects; there are many nonequilibrium effects involving line and plane defects, resulting from the fact that these regions (as sources and sinks for point defects) equilibrate more rapidly with environmental changes than does the bulk of the crystal. Accordingly, in systems which have not fully equilibrated, there can be considerable structural and compositional variations between the bulk of the crystal and the regions close to dislocations, grain boundaries, or the outer surfaces. See Bosman and van Daal [16] for examples.

A final point to note is that just as point defects, being variations from the perfect periodicity of the lattice, introduce defect levels into the electron energy-band scheme for the solid, so dislocations and surfaces introduce similar levels.

VII. DEFECTS IN GLASSES

There are two main difficulties in extending the treatment applied above for crystals to noncrystalline materials. First, since noncrystalline solids are not equilibrium structures, we cannot usually treat such "defects" as they contain as equilibrium species; in short, the mass-action approach is often inapplicable. It is true that where the noncrystalline material is a glass, quenched from some melt of the same composition, there is the possibility of studying the equilibrium structure of the melt and then projecting this into the structure of the glass (this is the same technique as that used to treat the effects of impurities in crystals when they are present in quasi-equilibrium, or to treat quenched crystalline materials — see Section V,A,5); also, when the glass may be regarded as the environment in which some reaction proceeds under equilibrium conditions, such as reducing the Pb ions in Pb glasses, the technique applied to crystal defects can be valid for that particular reaction. In general, however, the nonequilibrium nature of the noncrystalline solids restricts the use of the approaches discussed earlier in this paper.

A second problem is that a necessary preliminary step in using the concept of defects is the ability to define the perfect zero-defect structure; noncrystalline materials by their very nature are disordered and consequently selection of the zero-defect system is somewhat artificial.

Despite these problems, the idea of defects in glasses has found some use and two such cases may be described, the first showing an example of the approach suggested above, and the second concerning the effect of impurities on electronic conduction in noncrystalline systems.

For the silicate glasses, vitreous silica (SiO_2) is an obvious choice for the defect-free structure, since it consists entirely of a glass network containing Si_{Si}^x and O_O^x. (The notation of Kröger and

Vink [4], adapted where necessary, is used here to preserve consistency with the earlier sections; an alternative system has been proposed by Stevels [17], on whose work the following discussion is based.)

Impurity additions are of two types, those which enter the network substitutionally (formers) such as Al'_{Si}, and those which enter it interstitially (modifiers) such as $Na_i^{\cdot}$. Species are considered here to be in their conventional state of ionization. In addition we must recognize two types of oxygen site, the "perfect" type O_O^x (bridging oxygen), which is coordinated to two Si_{Si}^x, and the defect type, O'_* (nonbridging oxygen), which is directly coordinated to one Si_{Si}^x only. Then a number of possible neutrality approximations can be suggested, depending on the system composition; e.g.,

$$[Na_i^{\cdot}] = [Al'_{Si}] \tag{133}$$

$$[Na_i^{\cdot}] = [O'_*] \tag{134}$$

$$[Al'_{Si}] = [V_*^{\cdot}] \tag{135}$$

$$[Al'_{Si}] = 2[V_O^{\cdot\cdot}] \tag{136}$$

The definitions of sites are necessarily more vague in the present case than with crystals (the interstitial site in particular will probably have a number of configurations), but in other respects the effects of impurity addition can be treated in a very similar way. For electrical conductivity both $Na_i^{\cdot}$ and O'_* are known to be important; the first acts as an ionic charge carrier and the second breaks up the coherence of the network and thus eases the migration of the carriers. This feature is discussed in the paper by Stevels [17] and in Chapter 8.

For a second example we consider the semiconducting oxide glasses. Calculations of electron energy levels in noncrystalline systems show that in the basic features, such as the existence of energy bands and a forbidden gap, the energy schemes are similar to those of the crystalline counterparts, which in this case are the oxides. A distinguishing feature which arises from the deviations from periodic order in the glasses is the occurrence of band tailing; the deviations from order introduce electronic states in the noncrystalline form which are absent in the crystal and which, in the same way as the defect levels described earlier, lead to the existence of levels in the forbidden gap of the glasses. Since these defect states are in very high concentration in the noncrystalline solid they effectively form a band, and the result is that the original gap of the crystal is reduced in the disordered structure — a "tail of states" is added to the original band edges [see Fig. 20(a)].

The important result of this is that trace impurities, which, in a crystalline system, can greatly alter the energy-level structure by introducing defect levels close to the bands, are ineffective in noncrystalline systems because "defect" levels are already present in large concentration.

Effects of added "impurities" only become apparent when the additions are very substantial, and typically greater than 50 mole %, as with added V_2O_5 in the V_2O_5-P_2O_5 system. Under these circumstances the impurity levels are sufficiently dense to overlap, leading to a narrow impurity band within the original gap [see Fig. 20(b)]. Electronic conduction in glasses is discussed further in Chapter 8 and in the monograph by Owen [18].

VIII. DEFECTS AND ELECTRICAL CONDUCTIVITY

A. GENERAL

As indicated at the outset, the significance of a discussion of defects in the context of electrical conductivity arises from the fact that the defects are themselves the charge carriers. The conductivity is given by Eq. (1)

$$\sigma = \sum_a (n_a e_a \mu_a)$$

DEFECT STRUCTURE OF CERAMIC MATERIALS

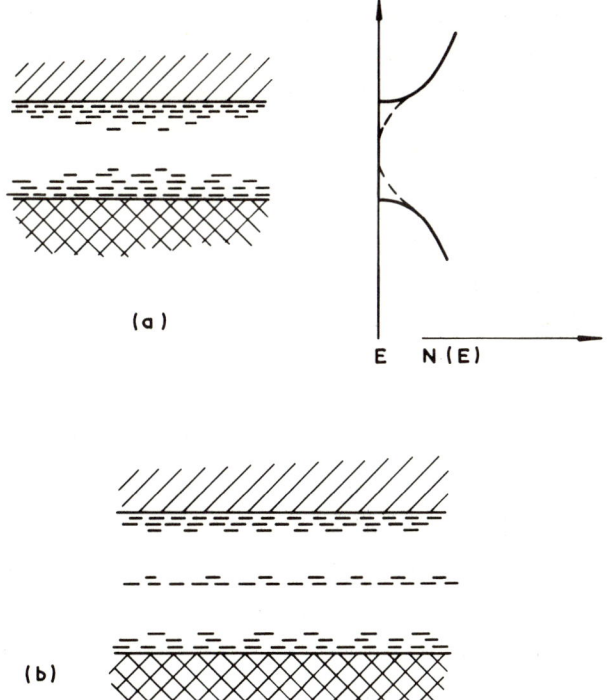

FIG. 20. (a) A schematic energy-band diagram for an amorphous solid, together with a plot of N(E), the number of electron states occurring at each energy E. The tail of states is shown as a dotted line. (b) A schematic energy-band diagram for amorphous P_2O_5-V_2O_5, showing the high density of vanadium levels in the band gap; the high density leads to formation of a narrow impurity band.

which can be split into the three major contributions to the electrical conductivity, namely, those arising from the ions, the electrons, and the electron holes,

$$\sigma = \sigma_i + \sigma_e + \sigma_h \tag{137}$$

$$= \sum_i (n_i z_i q \mu_i) - n_e q \mu_e + n_h q \mu_h \tag{138}$$

Here q is the magnitude of the electronic charge and z_i is the number of effective charges (±) carried by each of the ionic defect species i active in the system as carriers.

In the earlier sections we have been concerned solely with the concentration terms. The value of q is of course known; in the following we shall discuss very briefly the other terms in Eq. (138).

B. IONIC CONDUCTIVITY

Any of the effectively charged defects discussed earlier (excepting electrons and holes) gives rise to ionic conductivity if the defect can diffuse. Thus any of V_M'', V_M', $M_i^{\bullet\bullet}$, $M_i^{\bullet}$, $V_O^{\bullet\bullet}$, $V_O^{\bullet}$, O_i', O_i'' can act as carrier if they occur in MO. The mobility of the carrier is related to the tracer diffusion coefficient D_T, provided that the tracer and the defect move by the same mechanism. Then,

$$\sigma_i = n_i z_i q \mu_i = \frac{D_T z_i^2 q^2 n_i}{fkT} \qquad (139)$$

where f is a correlation factor (= 1 for interstitial diffusion, < 1 for vacancy diffusion). Values of mobility vary widely, but like the diffusion coefficient, mobility is an activated process, increasing strongly as the temperature is raised,

$$\mu_i = \frac{\mu_i^{\circ}}{T} \exp\left(-\frac{\Delta H_m}{RT}\right) \qquad (140)$$

Here, ΔH_m is the enthalpy of migration for the defect [21].

The form of Eq. (140) means that the graphs of defect concentrations vs temperature used earlier can easily be converted to conductivity vs temperature when the conductivity is ionic. Using Fig. 14, for example, and assuming the carrier to be V_M'' throughout ($\mu_{V_M''} > \mu_{V_O^{\bullet\bullet}}$), the conductivity plot appears as in Fig. 21.

The value of z_i for a given defect depends on the location of the defect level relative to the Fermi level as described earlier, (Section IV,D). For instance, from Fig. 11(b), depending on the oxygen pressure, the majority of the metal vacancies can exist as V_M'' or V_M', or V_M^x, where z_i would be -2, -1, or 0, respectively. It often happens in ionic systems that defects carry the charge

DEFECT STRUCTURE OF CERAMIC MATERIALS

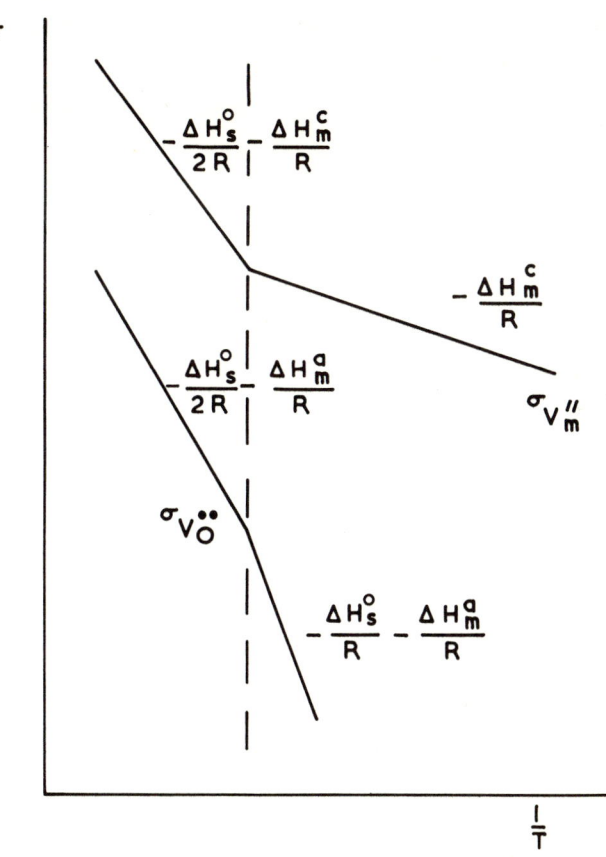

FIG. 21. The contributions to electrical conductivity of the defects shown in Fig. 14; $\mu_{V_M''} > \mu_{V_O^{\bullet\bullet}}$.

expected from that of the related ion (the defect levels lying close to the conduction and valence bands), but this cannot be assumed; z can be measured by concentration-cell measurements. The mobility of these conventionally charged defects is generally greater than that of abnormally charged ones.

C. ELECTRONIC CONDUCTIVITY

The mobility of electrons (and holes) depends on whether they are localized around particular sites in the crystal or are nonlocalized; as noted, this depends on whether the relevant electron band is narrow or broad.

For broad-band conduction the mobility is relatively large (> 1 cm^2/V-sec); above ambient temperatures it is reduced as the temperature is raised, since carrier scattering caused by thermal vibrations of the lattice atoms is enhanced. For broad-band semiconductors,

$$\mu \simeq T^{-m} \qquad (141)$$

where m is some number of the order unity, dependent on the dominant carrier-scattering mechanism.

For narrow-band conduction the mobility is small (< 1 cm^2/V-sec) and, at temperatures above $\theta/2$, where θ is the Debye temperature, it is an activated process, the carriers having to be excited from the localized sites where they are trapped by polarization of the lattice around them. The temperature dependence may then be represented by an equation similar to Eq. (140). The activation enthalpy for migration is typically small ($\sim$ 0.1 eV).

For both cases, we can note features which will enable us to construct conductivity vs activity diagrams from the concentration vs activity diagrams described earlier. In the first place, the mobility of electronic carriers is usually much higher than that of ionic carriers, even for the narrow-band conduction process, and secondly, the temperature dependence of the mobilities is less for electronic than for ionic carriers.

D. MIXED CONDUCTIVITY; CONDUCTIVITY VS OXYGEN PRESSURE DIAGRAMS

In many systems, depending on the conditions of temperature and activity, the overall conduction can be composed of substantial fractions of both ionic and electronic contributions — it is then termed mixed conductivity. From the indications above, the concentration of ionic carriers must be considerably larger than that of the electronic carriers in order to produce mixed conductivity and compensate for the mobility difference.

DEFECT STRUCTURE OF CERAMIC MATERIALS

To understand how the conductivity behavior can be developed from knowledge of the defect concentrations, we may consider Fig. 9(a). Assuming $\mu_{e'} > \mu_{h^\cdot} \gg \mu_{V_M''} > \mu_{V_O^{\cdot\cdot}}$, the contributions to the conductivity of the various defects are as shown in Fig. 22; the overall measured conductivity is electronic over the whole activity range for the situation shown.

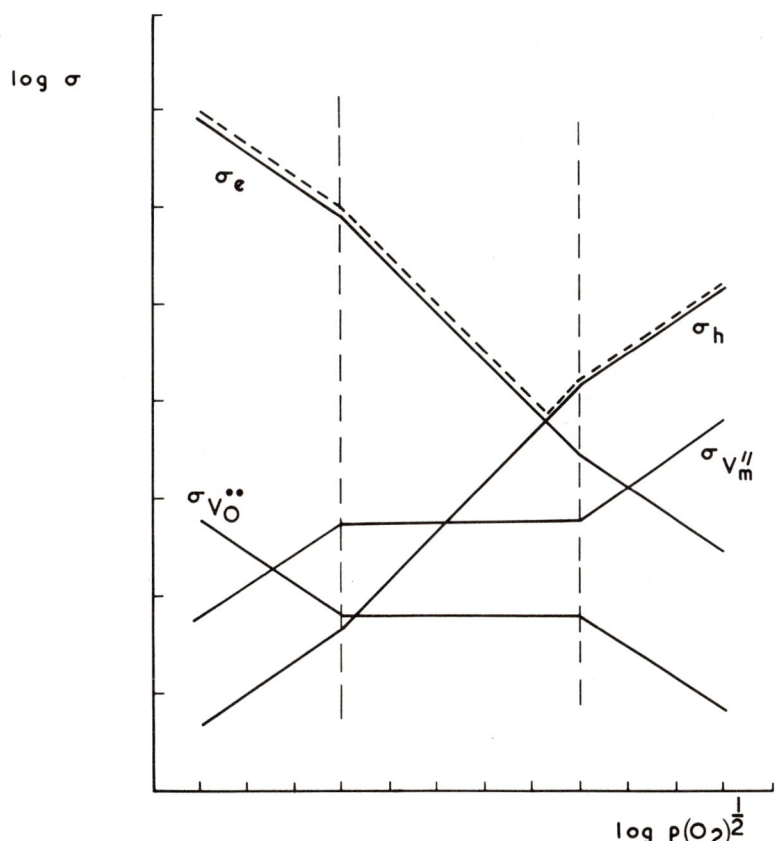

FIG. 22. The contributions to electrical conductivity of the defects shown in Fig. 9(a); $\mu_{e'} > \mu_{h^\cdot} > \mu_{V_M''} > \mu_{V_O^{\cdot\cdot}}$.

If the temperature is changed, then the form of this type of plot will also change owing to the different temperature dependences of the defect concentrations and mobilities. As an example, consider Fig. 10(a); the contributions to conductivity are shown in Fig. 23, and for the central range only in Fig. 24(a). If the temperature is lowered the concentration $[V_M'']$ is unchanged, since

$$2[V_M''] = [N_M^\bullet] = \text{const.}$$

and the ionic conductivity is given by

$$\sigma_{ion} \simeq \exp\left(-\frac{\Delta H_m}{RT}\right) \tag{142}$$

For the electronic contributions, assuming the mobilities are relatively temperature independent,

$$\sigma_e \simeq n = K_i \left(\frac{1}{2K_A}[N_M^\bullet]\right)^{1/2} p(O_2)^{-1/4}$$

$$= \text{const.}(N_M^\bullet, p(O_2)) \cdot \exp\left(-\frac{\Delta E_i}{RT} + \frac{\Delta H_A^O}{RT}\right)$$

and $\tag{143}$

$$\sigma_h \simeq p = \left(\frac{2K_A}{[N_M^\bullet]}\right)^{1/2} p(O_2)^{1/4}$$

$$= \text{const.}(N_M^\bullet, p(O_2)) \cdot \exp\left(-\frac{\Delta H_A^O}{RT}\right)$$

Taking $(\Delta E_i - \Delta H_A^O) \simeq \Delta H_A^O > \Delta H_m$, the central range at a lower temperature will appear as in Fig. 24(b), where a zone of ionic conduction has appeared between the two electronic zones. Note that these zones all appear within the one range, where there is the one neutrality regime $[N_M^\bullet] = 2[V_M'']$. A series of such figures taken at different temperatures will together give a three-dimensional plot of log [] vs log $p(O_2)^{1/2}$ vs 1/T (see Chapter 7), which contains all the information for the range.

In Fig. 24 the change in conductivity slope is caused by a change in charge carrier; the slope will also change on passing from one range to the next for a single-carrier type. Both cases are seen in Fig. 23.

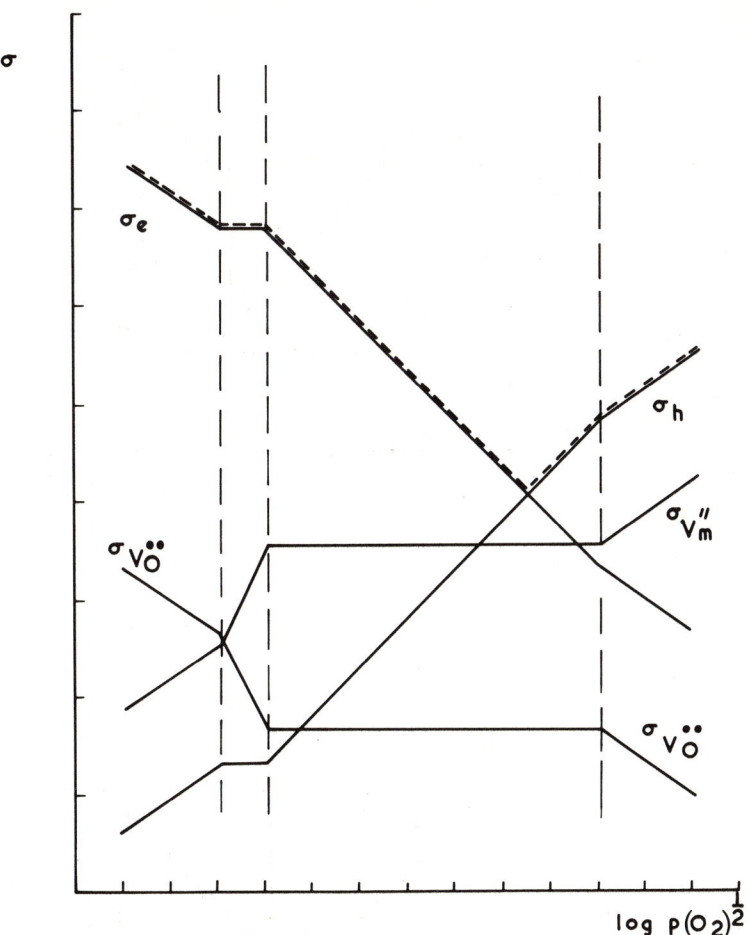

FIG. 23. The contributions to electrical conductivity of the defects shown in Fig. 10(a). The relative mobilities are as in Fig. 22.

E. INTERPRETATION OF CONDUCTIVITY MEASUREMENTS

All the discussion has so far been to the effect that if we understand the defect concentrations in a material, we can use knowledge of the defect mobilities to construct the conductivity behavior. In the experimental situation the problem is usually the other way round; we can make electrical measurements — can we then use these to construct the defect chemistry of the solid? Some

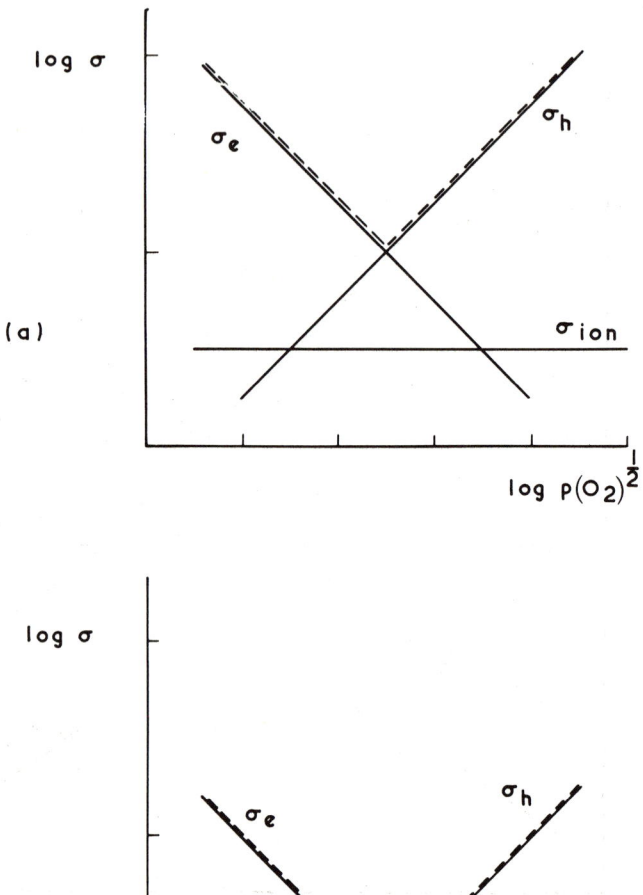

Fig. 24. A detail from Fig. 23 showing how changing the temperature can alter the relative electronic and ionic contributions to the total electrical conductivity in a mixed conductor; (b) shows the same range of log $p(O_2)$ as (a), but at a lower temperature. The relative mobilities are as in Fig. 22.

approaches and some examples are considered in this section. MO is again used as example.

DEFECT STRUCTURE OF CERAMIC MATERIALS   259

From the experimental electrical techniques discussed in Chapter 2, one can ideally find the following properties:

(a) ionic and electronic transport numbers (concentration cells, polarization studies)

(b) the number of charges carried by the majority ionic carrier (concentration cells)

(c) the concentration and sign of charge of the majority electronic carrier (measurements of conductivity and Hall or Seebeck effect)

These can all be measured as functions of temperature and the activity of the volatile component $[p(O_2)]$.

In conjunction with these electrical techniques there are important chemical techniques for studying defects; useful information is provided by tracer diffusion studies which can be used to identify the species of the majority ionic defect by comparison with ionic-conductivity measurements using Eq. (139). Thermobalance techniques, by measuring sample weight as a function of activity $[p(O_2)]$ can reveal how the majority atomic defect concentration changes with changing activity.

Generally, of course, all these data will not be available simultaneously, and the problem consists of using such data as are available to restrict the number of possible defect situations. A complicating factor is that one cannot assume that the defect behavior observed is that of the pure crystal; since small concentrations of impurity can dominate the neutrality condition (native defect concentrations being small), and since ceramic systems are difficult to purify, measurements on ostensibly pure systems are often representative of an impurity-controlled defect structure.

As examples of the treatment of conductivity data, we may consider some representative results.

1. <u>Conduction Is Electronic (or Ionic) and Varies with Activity $[p(O_2)]$, but No Bends Occur in the Diagram</u>

For this situation, one neutrality approximation applies in

the crystal throughout the experiments. Let us say conduction obeys

$$\sigma \simeq p(O_2)^{1/n} \qquad (144)$$

where n is a number. Then, making the approximation that the mobilities are independent of concentrations, we have that the charge-carrier concentration

$$[\ ] \simeq p(O_2)^{1/n} \qquad (145)$$

There are several explanations for this dependence, each representative of a different hypothesized neutrality approximation.

a. *The Carriers Appear in the Neutrality Approximation*. In this case we write the incorporation reaction for oxygen in terms of the carriers and the other species in the neutrality approximation. For example if the carriers are holes, and n in Eq. (145) is found to be 4, we can have

$$O_2 \rightarrow 2O_i' + 2h^\cdot \qquad (146)$$

or

$$O_2 \rightarrow 2O_O^x + 2V_M' + 2h^\cdot \qquad (147)$$

in order to get Eq. (145), the neutrality approximations being respectively $[O_i'] = p$ and $[V_M'] = p$. In general, for $\sigma \simeq p(O_2)^{1/n}$, where the carrier appears in the neutrality approximation, there will be n separate defects in the incorporation reaction (when it is written in terms of O_2) and these species will appear in the neutrality approximation.

As other examples, for MO_2, if holes are the carriers and

$$\sigma \simeq p(O_2)^{1/5} \qquad (148)$$

we could have

$$O_2 \rightarrow V_M'''' + 4h^\cdot + 2O_O^x \qquad (149)$$

with

$$p = 4[V_M''''] \qquad (150)$$

DEFECT STRUCTURE OF CERAMIC MATERIALS 261

For M_2O_3 with electrons as carriers and

$$\sigma \simeq p(O_2)^{-1/6} \qquad (151)$$

a possibility is

$$2O_O^x \rightarrow 2V_O^{\cdot\cdot} + 4e' + O_2 \qquad (152)$$

with

$$2[V_O^{\cdot\cdot}] = n \qquad (153)$$

For MO with ions (e.g., $V_O^{\cdot\cdot}$) as carriers and

$$\sigma \simeq p(O_2)^{1/6}$$

a possibility is

$$O_2 \rightarrow 2O_O + 2V_O^{\cdot\cdot} + 4V_M' \qquad (154)$$

with

$$2[V_O^{\cdot\cdot}] = [V_M'] \qquad (155)$$

b. <u>The Carriers Do Not Appear in the Neutrality Approximation</u>. In this case we write the incorporation reaction in terms of the carriers and the species in the neutrality approximation having the opposite charge. Then using the $p(O_2)$ dependence of this species we can find the relation between the carrier concentration and $p(O_2)$. For example, if we believe that the system displays intrinsic Schottky disorder, but that holes are the carriers, then Eq. (81) holds

$$[V_M''] = [V_O^{\cdot\cdot}] = \text{const.}(T)$$

and we can write reaction (70)

$$O_2 \rightarrow 2O_O^x + 2V_M'' + 4h^{\cdot}$$

whence

$$p \simeq p(O_2)^{1/4} \text{ since } [V_M''] \neq f[p(O_2)]$$

In general (see Fig. 12), any added charged species selects as compensating partner the defect from the neutrality condition with the opposite charge; in the present example holes can be considered as charged species introduced by the oxidation.

Another case where the carriers may not appear in the neutrality conditions is an impurity-dominated disorder; for example, if, with holes as carriers, we have Eq. (90)

$$[N_M^\bullet] = 2[V_M''] = \text{const.}$$

then Eq. (70) and $p \simeq p(O_2)^{1/4}$ would again apply.

From the examples it is apparent that a number of possible solutions to explain a dependence such as Eq. (144) exist; choice between these possibilities is made on the basis of the likely physical solutions and of other available measurements, such as the level of impurity in the sample, or diffusion data or thermobalance data.

A useful technique for identifying the species in the neutrality approximation is to add dopants of known effective charge, and then to determine the compensating native defect (Fig. 12).

2. Conduction Is Electronic (or Ionic) and Is Independent of Activity

Then if the charge carrier appears in the neutrality approximation the system is either one of intrinsic electronic (or ionic) disorder, or is impurity controlled. Varying the impurity concentration can distinguish between the two alternatives, since only an impurity-controlled situation would be affected. One would also look for variation in the temperature 1/T curve, such as the appearance of an extrinsic region at lower temperature or of an intrinsic region at higher temperature for the respective situations. If the carrier is not in the neutrality approximation then as in Section VIII,E,1 above, there are many possibilities (e.g., as in Fig. 11(b), right-hand side, with V_M'' as the carrier).

3. More than One Linear Region Appears in the Activity Graph

This may result from a change from one carrier type to another within one neutrality range (Fig. 24) or from a single carrier type crossing from one neutrality range to another. Measurements of transport number are clearly important for interpreting the mixed-carrier situations; when the individual carrier dependences are derived, the techniques in Section VIII,E,1 and 2 above can be applied.

4. Conduction Varies with Impurity Content

The impurity then occurs in the neutrality approximations; the charge carrier may or may not. As an example of what may happen, the case of Fig. 12(b) is shown in Fig. 25 assuming $\mu_{h\cdot} > \mu_e \gg \mu_{V_M''}$.

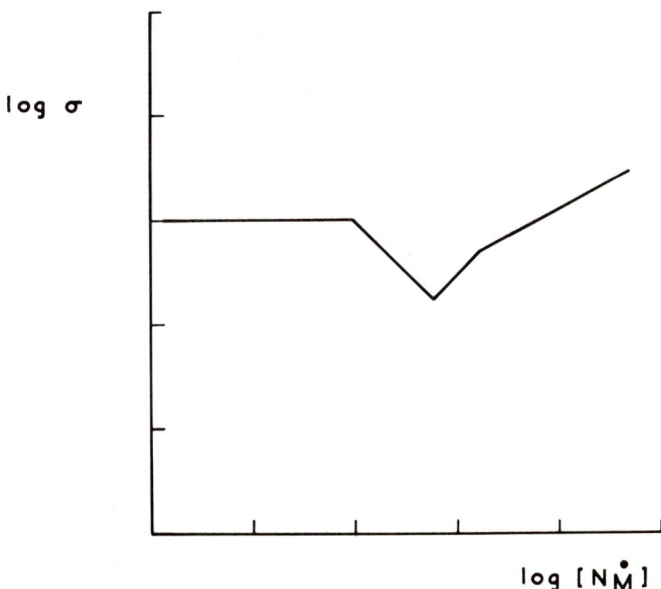

FIG. 25. The effect of impurity addition on electrical conductivity; the defect structure is that of Fig. 12(b); $\mu_{h\cdot} > \mu_{e'}$.

5. Conduction Varies with Temperature

This of course is the most common experiment and by itself throws little light on the defect chemistry. If several ranges appear on the log σ vs $1/T$ plot then it may be possible to suggest extrinsic and intrinsic ranges or extrinsic and association ranges from the relative slopes of the two lines. If it is to be informative this technique requires combination with one of the other methods, such as impurity variation or activity variation, or if necessary both.

IX. CONCLUSION

In this chapter we have discussed point defects, how they arise, how they influence the measured electrical conductivity, and how the conductivity measurements can in turn be used to identify them. Emphasis has been placed throughout on defects as chemical species whose behavior can be treated by the methods of chemical equilibrium.

The discussion has been in terms of an unspecified oxide MO and of hypothetical defect situations; this has allowed us to concentrate more readily on the language of defect chemistry and on the various methods for treating the effects of the different experimental variables, temperature, impurity level, and nonstoichiometry on defect concentrations.

The application of defect chemistry to real systems follows closely that outlined in the hypothetical situations described earlier; there are examples elsewhere in this book, and a brief, and by no means comprehensive, listing of some instances from the literature involving point defect studies in real oxide systems, is presented in the Appendix. A sampling of these works should emphasize that a familiarity with the principles of defect chemistry is basic to the interpretation of electrical-conduction phenomena in ceramics.

There are few instances where our knowledge of the defect chemistry of a ceramic material is in any way complete; in those cases experimental determination of electrical conductivity has been one of the most important contributions to the knowledge. Consequently one of the main purposes of conductivity measurements is to reveal further the defect structures of these materials so that the many other properties that are determined by defects may be better understood.

In a complementary sense, the performance and interpretation of electrical measurements requires an understanding of the role of defects and is helped even at early stages by attempts to find hypothetical solutions for the defect structure; this structure is a convenient and concise form in which all the relevant experimental data can be summarized, and it suggests at each stage where further experimental effort is most likely to be rewarding.

APPENDIX

The following is a list of papers discussing the application of defect chemistry to some specific systems.

Al_2O_3	R. J. Brook, J. Amer. Ceram. Soc., 55, 114 (1972).
BeO	C. F. Cline and H. W. Newkirk, J. Chem. Phys., 49, 3496 (1968).
CeO_2	P. Kofstad and A. Z. Hed, J. Amer. Ceram. Soc., 50, 681 (1967).
CoO	N. G. Eror and J. B. Wagner, Jr., J. Phys. Chem. Solids, 29, 1597 (1968).
FeO	P. Kofstad and A. Z. Hed, J. Electrochem. Soc., 115, 102 (1968).
$La_2O_3(CaO)$	T. H. Etsell and S. N. Flengass, J. Electrochem. Soc., 116, 771 (1969).
MnO	A. Z. Hed and D. S. Tannhauser, J. Chem. Phys., 47, 2090 (1967).
NiO	I. Bransky and N. M. Tallan, J. Chem. Phys., 49, 1243 (1968).

SrO	W. D. Copeland and R. A. Swalin, *J. Phys. Chem. Solids*, 29, 313 (1968).
TiO_2	R. N. Blumenthal, J. Baukus, and W. M. Hirthe, *J. Electrochem. Soc.*, 114, 172 (1967).
V_2O_5	T. Allersma, R. Hakim, T. N. Kennedy, and J. D. Mackenzie, *J. Chem. Phys.*, 46, 154 (1967).
Y_2O_3	N. M. Tallan and R. W. Vest, *J. Amer. Ceram. Soc.*, 49, 401 (1966).
$ZrO_2(CaO)$	J. W. Patterson, E. C. Bogren, and R. A. Rapp, *J. Electrochem. Soc.*, 114, 752 (1967).

REFERENCES

1. C. Wagner and W. Schottky, *Z. Phys. Chem. (Leipzig)*, B11, 163 (1931).

2. K. Denbigh, *The Principles of Chemical Equilibrium*, Cambridge Univ. Press, Cambridge, England, 1966.

3. F. A. Kröger, *The Chemistry of Imperfect Crystals*, North-Holland, Amsterdam, 1964.

4. F. A. Kröger and H. J. Vink, in *Solid State Physics*, Vol. 3 (F. Seitz and D. Turnbull, eds.), Academic Press, New York, 1956, p. 307.

5. W. van Gool, *Principles of Defect Chemistry of Crystalline Solids*, Academic Press, New York, 1966.

6. L. Eyring and M. O'Keeffe, *The Chemistry of Extended Defects in Non-Metallic Solids*, North-Holland, Amsterdam, 1970.

7. F. R. N. Nabarro, *Theory of Crystal Dislocations*, Oxford Univ. Press, Oxford, England, 1967.

8. L. T. Chadderton, *Radiation Damage in Crystals*, Methuen, London, 1965.

9. L. E. Orgel, *An Introduction to Transition-Metal Chemistry*, Methuen, London, 1960.

10. F. A. Kröger, *J. Phys. Chem. Solids*, 29, 1889 (1968).

11. J. B. Goodenough, in *Progress in Solid State Chemistry*, Vol. 5 (H. Reiss, ed.), Pergamon, Oxford, 1971.

12. G. Brouwer, Philips Res. Rep., 9, 366 (1954).

13. H. Schmalzried, in Progress in Solid State Chemistry, Vol. 2 (H. Reiss, ed.), Pergamon, Oxford, 1965.

14. G. G. Libowitz, in Mass Transport in Oxides (J. B. Wachtman, Jr. and A. D. Franklin, eds.), NBS Special Pub. 296, U. S. Govt. Printing Office, Washington, D. C., 1968, p. 109.

15. A. B. Lidiard, in Handbuch der Physik, Vol. 20, Springer-Verlag, Berlin, 1957, p. 246.

16. A. J. Bosman and H. J. van Daal, Advan. Phys., 19, 1 (1970).

17. J. M. Stevels, in Handbuch der Physik, Vol. 20, Springer-Verlag, Berlin, 1957, p. 350.

18. A. E. Owen, Contemp. Phys., 11, 226 (1970).

19. S. van Houten, J. Phys. Chem. Solids, 17, 7 (1960).

20. A. J. Dekker, Solid State Physics, Prentice-Hall, Englewood Cliffs, New Jersey, 1900 (1957).

21. L. W. Barr and A. B. Lidiard, in Physical Chemistry: an Advanced Treatise, Vol. 10, Solid State (W. Jost, ed.), Academic Press, New York, 1970, p. 151.

Chapter 4

ELECTRONIC CONDUCTION

J. M. Wimmer and I. Bransky

Aerospace Research Laboratories (LL)
Metallurgy and Ceramics Research Laboratory
Wright-Patterson Air Force Base, Ohio

I.	INTRODUCTION	270
II.	THEORETICAL FORMS OF THE DRIFT MOBILITY	271
	A. Lattice Scattering	271
	B. Ionized Impurity Scattering	272
	C. Polaron Mobility	272
III.	CHARGE-CARRIER CONCENTRATION	279
	A. Normal Ionic and Transition-Metal Compounds	279
	B. Departure from Stoichiometry	282
	C. Doping	284
	D. Activation Energy	286
IV.	EXPERIMENTAL DETERMINATION OF THE CHARGE-CARRIER CONCENTRATION AND DRIFT MOBILITY	287
	A. Departure from Stoichiometry	288
	B. Seebeck Coefficient	290
V.	ANALYSIS OF ELECTRICAL-CONDUCTIVITY AND THERMAL emf MEASUREMENTS FOR TWO TYPES OF CHARGE CARRIERS . . .	293
VI.	EXAMPLES IN CERAMIC MATERIALS	296
	A. Al_2O_3, MgO, BeO	298
	B. Iron-Group Monoxides	301
	C. ZrO_2, HfO_2, ThO_2	304
	D. TiO_{2-x}, Nb_2O_{5-x}, CeO_{2-x}, and UO_{2+x}	306
	REFERENCES	307

I. INTRODUCTION

A flux of particles, J_p, can be expressed as the product of particle concentration c and velocity. This velocity can, in turn, be written as the product of the force acting on the particle and the particle mobility B, and as a result the units of particle mobility are velocity per unit force. In the case of electron or hole motion in an electric field E, the magnitude of the force is eE, where e is the charge on an electron, and in this case it is customary to define a new mobility μ, which has the units of drift velocity per unit electric field. The flux of charge, J_c, is therefore given by

$$J_c = |e| J_p = |e| c \mu E = \sigma E \qquad (1)$$

where σ is the electronic conductivity, which varies from material to material according to the magnitude of the concentration-mobility product. In general, since both electrons and electron holes contribute to conduction σ is written in terms of the partial conductivities of electrons, σ_n, and holes, σ_p, as

$$\begin{aligned}\sigma &= \sigma_n + \sigma_p \\ &= |e| n\mu_n + |e| p \mu_p \end{aligned} \qquad (2)$$

where n and p are the electron and hole concentrations, respectively. If the concentration is expressed in cm^{-3}, the charge in coulombs, and the mobility in $cm^2/V\text{-sec}$, then the units of σ are $ohm^{-1}\text{-}cm^{-1}$. In general the mobility is a tensor, but in the case of isotropic materials or for simple spherical energy surfaces and isotropic scattering it is a scalar.

It is clear from Eq. (1) that studies of the electrical conductivity yield information only about the product of charge-carrier concentration and drift mobility. The objective of this chapter will be to discuss briefly the natures of c and μ individually, to present several combinations of measurements which are commonly used to separate c and μ in metal oxides, and finally to examine the behavior of the electrical conductivity of some typical oxides of importance to the ceramist. We will proceed by discussing first some of the most important theoretical forms of the drift mobility in nonpolar broad-band and in polar narrow-band semiconductors, since the specific form of the mobility observed is of particular interest in attempting to distinguish between alternative electronic conduction mechanisms in ceramic materials.

II. THEORETICAL FORMS OF THE DRIFT MOBILITY

A. LATTICE SCATTERING

In the case of pure nonpolar broad-band semiconductors the drift mobility is predominantly determined by acoustic phonon scattering of the electrons. One method for the calculation of acoustic-mode scattering is that of the "deformation potential", originally used by Bardeen and Shockley [1]. The conduction and valence bands result from the interaction of atomic energy levels that occurs as the atoms approach one another during the formation of the solid crystal. The position of the band edges is, therefore, dependent on the lattice parameter of the crystal, and as the lattice atoms vibrate about their equilibrium positions their vibrational motion causes a local variation in the energy of the band edges. This shift in the band edge, δE, is given by $\delta E = \epsilon_1 \Delta$, where Δ is the dilatation, or fractional change in volume, and ϵ_1 is a constant called the deformation potential constant. These fluctuations in band energy cause an abrupt change in the potential energy of the electron (or hole) moving in the band. Since energy is conserved, the kinetic energy of the carrier also changes abruptly in this process and the carrier is, in effect, scattered.

The drift mobility derived by this method has the form

$$\mu = \frac{(8\pi)^{1/2} \hbar^4 c_\ell \rho}{3(kT)^{3/2} m^{*5/2} \epsilon_1^2}$$

$$= (\text{const.}) \, T^{-3/2} \tag{3}$$

where c_ℓ is the average velocity for longitudinal acoustic phonons and ρ is the density of the crystal.

B. IONIZED IMPURITY SCATTERING

In impure semiconductors ionized donor and acceptor centers are positively and negatively charged, respectively, and will, therefore, serve as scattering sites, which tend to limit the electronic mobility. The scattering potential is taken as the coulombic potential in a continuous dielectric medium, and the calculation is similar to that for Rutherford scattering. Conwell and Weisskopf [2] calculated an approximation for the relaxation time τ for this case and showed that in the case of ionized impurity scattering the mobility is given by

$$\mu = \frac{8(2)^{1/2} \kappa^2 (kT)^{3/2}}{\pi^{3/2} N_i \, e^3 (m^*)^{1/2}} \ln \left[1 + \left(\frac{3\kappa kT}{e^2 N_i^{1/3}} \right)^2 \right]^{-1} \tag{4}$$

where κ is the dielectric constant and N_i is the concentration of charged impurities. Impurity scattering, therefore, leads to a drift mobility which varies approximately as $T^{3/2}$, in contrast to the $T^{-3/2}$ dependence obtained in the case of lattice scattering.

C. POLARON MOBILITY

In compounds with predominantly ionic character the mobility of an electron is determined to a large extent by its interaction with the polar modes of the crystal. The self-trapping of an electron in an ionic crystal, first suggested by Landau [3], was one of the earliest approaches to a polaron theory. The electron was considered to interact coulombically with the ions, producing a potential well surrounding the electron which was then self-trapped (or bound) within it. Since the ions are much heavier than the electron they do not return to their equilibrium positions during a half-period of the electronic motion within the trapping potential, and the electron is effectively localized. The electron and its surrounding polarization cloud is commonly referred to as a polaron. A number of theories have been developed more recently which treat in detail the electrical, optical, and galvanomagnetic properties expected for large and small polarons. A comprehensive review of the subject is given by Appel [4].

Present polaron theories predict that the energy of an electron in an ionic semiconductor is lowered relative to its energy in a rigid lattice by the amount W_p, the polaron energy, because of the electron-phonon interaction, and that the effective mass of the electron is increased to the value m_p, the polaron mass, which is greater than m^*, the effective mass of the electron in a rigid lattice. Two extreme cases can be considered. For each, the appropriate formulas for the electronic behavior depend on the coupling constant α_c given by

$$\alpha_c^2 = \tfrac{1}{2} \frac{e^2/\kappa_p r_p}{\hbar \omega_o} \tag{5}$$

where ω_o is the mean frequency of the optical phonons, r_p is the polaron radius, given by

$$r_p = \frac{\kappa_p \hbar^2}{m^* e^2} \tag{6}$$

and

$$\frac{1}{\kappa_p} = \frac{1}{\kappa_\infty} - \frac{1}{\kappa_s} \tag{7}$$

where κ_s and κ_∞ are the static and high-frequency dielectric constants. The two extreme cases are (a) $\alpha_c < 1$, weak coupling, which gives rise to the formation of large-radius polarons moving in a band with an effective mass greater than that of an electron in a rigid lattice, and (b) $\alpha_c \gg 1$, strong coupling, which gives rise to the formation of small-radius polarons with r_p of the order of the interionic distance.

1. The Large-Polaron Case

Theories describing the behavior of large polarons generally treat the polarization field as a continuum, since the lattice distortion extends over several lattice constants (an approach which clearly becomes unrealistic in the case of strong coupling, where the distortion is restricted to the vicinity of the electron). For small values of α_c, Howarth and Sondheimer [5] derived the form of the large-polaron mobility under the assumption of a simple band with a parabolic energy dependence and the assumption that the longitudinal optical phonons all have the same frequency. The mobility, as reformulated by Petritz and Scanlon [6] is given by

$$\mu = \frac{e}{2m^* \omega_\ell \alpha_c} \frac{8}{3\pi^{1/2}} \chi(z) \frac{\exp(z-1)}{z^{1/2}} \qquad (8)$$

where ω_ℓ is the angular frequency of a longitudinal optical phonon, $z = \theta_D/T$, where θ_D is the Debye temperature of the optical branch of the phonon spectrum, and $\chi(z)$ is given, for low temperatures or large values of z, by $\chi(z) = (\text{const.}) z^{1/2}$. Under these conditions the temperature dependence is contained in the exponential term. Other forms for the drift mobility have been derived within the large-polaron framework [7,8,9]. The predominance of an $\exp(\theta_D/T)$ dependence for the drift mobility at low temperatures is common to all of these large-polaron theories.

2. The Small-Polaron Case

The preceding description of large-polaron conduction is based essentially within the framework of the band model. It can be shown [10,11], however, that there is a lower limiting value of the drift mobility for consistency with the band scheme. The limit depends on the band width compared with kT and on the criteria used, but a lower limit of about 0.5 cm^2/V-sec is representative. For materials with lower mobilities a new description must be sought.

In the case of ionic semiconductors in which the conduction band is sufficiently narrow and the electron-lattice interaction sufficiently strong the small-polaron model can be applied. The electron now moves so slowly that it is considered to be self-trapped in its own induced polarization field generated by the ionic lattice, and therefore to be localized at a given lattice site. Two different transport mechanisms are predicted by small-polaron theories [12,13]: (a) tunneling behavior, or small-polaron band conduction, described by a wave-like expansion of the localized states, and (b) hopping of the small polaron between adjacent sites. Although earlier small-polaron theories predicted the transition between these mechanisms to occur at a temperature of the order of $\theta_D/2$ with small-polaron band conduction predominating at low temperatures, recent arguments suggest lower and lower values for the transition temperature. Also, if the fluctuations in potential energy between neighboring sites caused by impurities exceeds the band width then the band character can be destroyed. On the basis of these arguments, Bosman and van Daal [11] question whether small-polaron band conduction will ever be observed in practice.

In the case of the second transport mechanism, small-polaron hopping, the exchange of sites is accompanied by the absorption or emission of a number of optical phonons. If the ionic crystal is taken to be a continuous dielectric medium [14,15,16], then the small-polaron energy is roughly approximated by

$$W_p = \tfrac{1}{2}e^2/\kappa_p r_p \tag{9}$$

and it is clear from Eq. (5) that the requirement of a strong coupling constant for small-polaron formation corresponds to the polaron energy being much larger than the mean phonon energy. If we introduce J, the resonance or electron-transfer integral (in the tight-binding approach to the band structure of a one-dimensional lattice, the band width ΔW is related to the resonance integral by the relation $\Delta W = 4J$), then the nonadiabatic small-polaron case is characterized by $J < \frac{1}{2}W_p$. In this case, where the probability of an electron tunneling from one site to the next during an excited state of the system is small, the drift mobility is predicted to be thermally activated with the form [12]

$$\mu = \frac{e\pi^{1/2}}{2\hbar} \frac{J^2 a^2}{W_h^{1/2}} (kT)^{-3/2} \exp\left(-\frac{W_h}{kT}\right) \qquad (10)$$

and the hopping energy is given to a good approximation by

$$W_h = \tfrac{1}{2} W_p \qquad (11)$$

In the adiabatic small-polaron case ($J \simeq \tfrac{1}{2} W_p$) the electron can tunnel back and forth between sites several times during one excited state of the system, and the thermally activated hopping of the charge carrier is slightly modified. Emin and Holstein [16] have shown that in this case the drift mobility is given by

$$\mu = \frac{ea^2 \omega_o}{kT} \exp\left(-\frac{W_h'}{kT}\right) \qquad (12)$$

where $W_h' = \tfrac{1}{2} W_p - J$. The reduction of the hopping energy by the factor J is not unique and other reductions have been suggested [17] for the adiabatic case.

Although earlier theories [12] for the nonadiabatic small polaron at high temperatures predicted that $W_h = \tfrac{1}{2} W_p$, recent refinements indicate that even in this case W_h may be somewhat lower or higher than $\tfrac{1}{2} W_p$. Mott [18] has shown that if large numbers of polaron centers are present and the potential wells are close enough for the polarization clouds to overlap, then W_h is decreased by the factor $1 - r_p/R$, where R is the distance between the centers, and the measured value of the hopping energy is given by

$$W_h = \tfrac{1}{2} W_p \left(1 - \frac{r_p}{R}\right) \qquad (13)$$

On the other hand, an increase in the hopping energy in highly doped or highly defective crystals may be observed because of Anderson localization [19]. This type of localization is a result not of electron-phonon interaction but of random fields arising from the presence of point defects or a disordered arrangement of the atoms. The typical case, treated by Anderson [19], is impurity conduction. If two such localized centers have quantized energy levels differing by W_D, the average microfield fluctuation, then the hopping probability of an electron between the centers is proportional to $\exp(-W_D/kT)$. Therefore, in the case of small-polaron hopping between sites which differ in energy by W_D because of random fields, the overall activation energy measured for hopping is given by [17] $W_h + \frac{1}{2}W_D + W_D^2/16W_h$, where the last term is small because W_h is expected to be considerably greater than W_D. Still another kind of localization which affects the validity of the approximation in Eq. (11) is spin-state localization, which was postulated by Jonker [20] to explain the activated mobility of electrons in $LaCoO_3$. This localization is associated with the difference in spin states of the different valence states of Co in the oxide, and arises because an internal spin rearrangement is required at a site before a hopping transition can occur.

The most frequently proposed candidates for small-polaron hopping conduction are those oxides containing transition-metal cations, e.g., NiO, CoO, MnO, TiO_2, Cr_2O_3, α-Fe_2O_3, Fe_3O_4, $BaTiO_3$, UO_2. The basic argument offered in early studies [21,22] in support of a small thermally activated mobility in the d-bands of these transition-metal oxides was based on the ionic character of these materials as well as on the small overlap of the partially filled d-orbitals. These arguments were based on considerations of the relatively large separation (about 4 Å in the Fe-group monoxides) between the transition-metal cations and the relatively high and wide potential barriers of the oxygen anions, which were assumed to impede charge transfer between the cations.

Recently, however, doubts have arisen about the applicability of the small-polaron hopping mechanism to most of the transition-metal oxides. These doubts have been based mainly on carefully taken experimental measurements [11,23-25] which have failed to show the characteristic activated dc mobility at high temperatures. These experiments, which will be discussed in some detail later, cannot, however, really rule out the validity of the hopping process in these oxides. The activation energy for hopping in these materials is in fact expected to be very small and difficult to observe experimentally. A crude estimate of W_h can be obtained from Eqs. (9) and (11). If we use $\kappa_p \simeq 7$ and $r_p \simeq 4$ Å, values which are reasonable for the transition-metal monoxides of the iron group, then $W_p \simeq 0.25$ eV and, if the charge carriers are strongly localized in these materials, we should expect to observe hopping energies of the order of 0.1 eV. In the adiabatic small-polaron case [16], where $\frac{1}{2}W_p \simeq J$, the hopping energy W_h' appearing in Eq. (12), given by $W_h' = \frac{1}{2}W_p - J$, would be expected to be very small. The temperature dependence of the drift mobility of a small polaron in the adiabatic case would, therefore, in principle be less characteristic of a thermally activated hopping process.

In both Eqs. (10) and (12), the temperature dependence of the drift mobility is further reduced by an inverse temperature dependence in the pre-exponential factor. Therefore not only is the hopping energy expected to be small in these materials, but its evaluation is made still more difficult and uncertain by the relative importance of the form of the pre-exponential factor that is used in the interpretation of the data. The combinations of experimental measurements currently in use for the determination of the drift mobility in metal oxides are unfortunately not really sensitive or accurate enough for the study of processes with activation energies on the order of kT.

ELECTRONIC CONDUCTION   279

III. CHARGE-CARRIER CONCENTRATION

The conduction in n- or p-type semiconductors is usually categorized according to the origin of the free charge-carrier concentration, intrinsic conduction arising from excitation across the band gap and extrinsic conduction arising from excitation from localized states within the band gap. For a given semiconductor the type observed will depend on the concentration of point defects, line defects, impurities, and the temperature. Rarely are ceramic materials free enough from all kinds of defects and impurities to exhibit intrinsic behavior over a large temperature range, and in practice there is more interest in the extrinsic or defect-controlled behavior. The band scheme and defect structure of ceramic semiconductors are discussed to a large extent in Chapters 1 and 3, so only a few of the features relevant to the charge transport properties will be discussed here.

A. NORMAL IONIC AND TRANSITION-METAL COMPOUNDS

Consider the ionic (polar) compounds with the formula $M^{2+}X^{2-}$. Kroger [26] has pointed out that such compounds can be divided into two groups — the normal ionic compounds in which the cations have a rare-gas configuration (e.g., MgO, NaCl), and the transition-metal compounds in which the cations have unfilled inner shells (e.g., monoxides of Fe, Ni, Co, Mn). It has been suggested [27] that in the $M^{2+}X^{2-}$ compounds, particularly in those having narrow transport bands, the conduction band can be viewed as a combination of states which, when occupied, correspond to electrons residing predominantly on M ions (creating effectively M^{+} ions). The conduction band can, therefore, be called the M^{+} band, while for analogous reasons the valence band in the normal compound can be called the X^{2-} band, i.e., the charge state when occupied by an electron. For the normal $M^{2+}X^{2-}$ compound the position of the middle of the conduction and valence bands can be estimated from [26,27]

$$E_c = -I_{M^+} + B \frac{q^2}{r_{MX}} + \alpha_p \tag{14}$$

$$E_v = A_{X^-} - B \frac{q^2}{r_{MX}} + \beta_p \tag{15}$$

where I_{M^+} is the second ionization energy of M, A_{X^-} the electron affinity of X^-, B the Madelung constant, r_{MX} the shortest M-X distance, and α_p and β_p are lattice polarization energies. In the transition-metal compound the X^{2-} band (the upper filled band in the normal compound) is expected to be at about the same level as in the normal compound, but the position of the conduction band is given by

$$E'_c = I_{M^+} + B \frac{q^2}{r_{MX}} + \alpha_p + C_+ - C_{2+} \tag{16}$$

where C_+ and C_{2+} are the crystal-field stabilizations of M^+ and M^{2+} ions. In the transition-metal compound the upper filled band lies above the X^{2-} band and is due to M^{2+} states, i.e., filled M^{3+} states. In the normal compound, the third ionization energy of M is very large, since this electron is being removed from a completely filled outer shell, and therefore the M^{2+} band lies well below the X^{2-} band. In the transition-metal compound, however, the removal of an electron from the unfilled outer shell is relatively easy, and as a result the M^{2+} band lies above the X^{2-} band. For example, in the iron-group monoxides the position of this band is 5-8 eV above the O^{2-} band [26,27]. Both the M^{2+} and M^+ bands are usually very narrow in these monoxides because of the small overlap of the 3d orbitals. A diagram of the band scheme for the two types of compounds is shown in Fig. 1. This diagram applies to the situation at high temperatures where no spin correlation exists. It is apparent from the diagram [26] that it is energetically easier to incorporate an excess of X in the transition-metal compound than in the normal compound, since the addition of X in the normal compound introduces two holes in the X^{2-} band (or in the localized levels V'_M and V''_M which lie just above X^{2-}), while in the transition-metal compound, the holes are introduced into the high-lying M^{2+} level (or in the V'_M and V''_M levels just above M^{2+}).

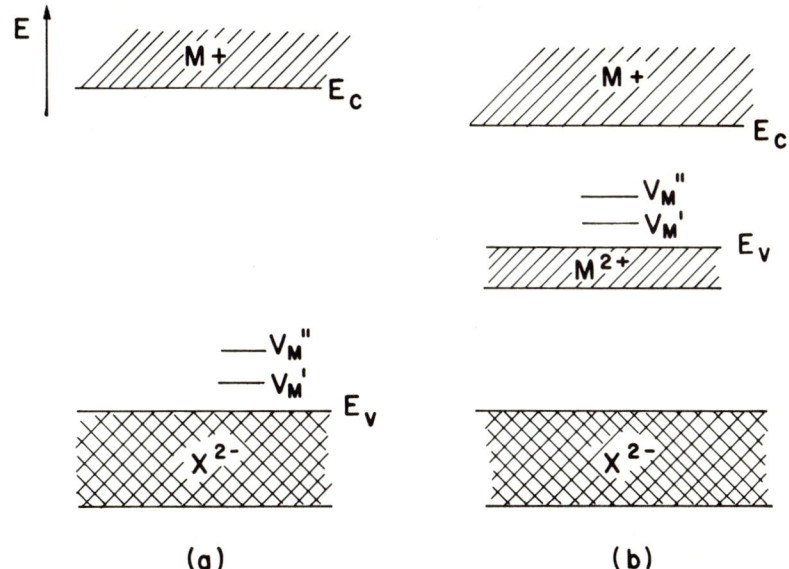

FIG. 1. Energy-band scheme for (a) a normal $M^{2+}X^{2-}$ compound and (b) a transition-metal compound (after Kroger [26]).

The transition-metal compound is, therefore, an example of the controlled-valence principle [28].

Free charge carriers can be created intrinsically by excitation across the M^{2+}-M^+ (or X^{2-}-M^+) energy-band gap or extrinsically from localized point-defect states which lie within the forbidden band. Due to the large energy gap found in most ceramic semiconductors, and the lack of sufficient purity, intrinsic conduction is rarely observed. There are exceptions; e.g., $LaCoO_3$ has a narrow band gap of 0.3 eV [20], and intrinsic conduction is observed at relatively low temperature. The relative ease of introducing charge carriers into transition-metal compounds by the principle of controlled valence, either by doping with foreign atoms or by departure from stoichiometry, is used extensively to study the defect structure through the variation of the charge-carrier concentration. Although the subject of nonstoichiometry and its pressure and temperature dependence is discussed in Chapter 3, it is instructive to analyze one case to demonstrate the factors which control the concentration of charge carriers due to a departure from stoichiometry.

B. DEPARTURE FROM STOICHIOMETRY

Consider the following defect-structure reactions involving metal vacancies and holes:

$$\tfrac{1}{2}O_{2(g)} \rightleftharpoons O_O + V_M^x \qquad (17)$$

$$V_M^x \rightleftharpoons V_M' + h^\cdot \qquad (18)$$

$$V_M' \rightleftharpoons V_M'' + h^\cdot \qquad (19)$$

These reactions can occur at high temperature, e.g., in a monoxide such as CoO, MnO, or NiO. The departure from stoichiometry arises due to the incorporation of oxygen atoms from the gas phase, $\tfrac{1}{2}O_{2(g)}$, onto the normal lattice sites, O_O. The excess oxygen is accompanied by a cation vacancy V_M^x which can, according to Eqs. (18) and (19), produce holes and become either singly or doubly ionized. As discussed above, these holes will move in a valence band consisting of M^{2+} states. As long as the concentration of vacancies and holes is sufficiently small the mass-action law can be applied, with the result that

$$[V_M'] = \frac{K_{V_M'} P_{O_2}^{1/2}}{p} \qquad (20)$$

and

$$[V_M''] = \frac{K_{V_M''} P_{O_2}^{1/2}}{p^2} \qquad (21)$$

where p is the concentration of holes and the brackets denote concentration (usually in mole fraction). In the low defect-concentration range, which in this model occurs at low oxygen partial pressure, it is reasonable to assume that impurities will influence behavior. For simplicity it will be assumed that the effective impurity concentration is independent of temperature and oxygen pressure and that the charge state relative to the lattice is ±1 electronic charge. If it is assumed that the crystal as a whole maintains charge neutrality then

$$p = 2[V_M''] + [V_M'] \pm [F] \qquad (22)$$

where [F] is the concentration of impurities. Inserting Eqs. (20) and (21) into Eq. (22), the following relation between p and P_{O_2} is obtained,

$$p^3 \mp [F] p^2 - K_{V_M'} p P_{O_2}^{1/2} - 2K_{V_M''} P_{O_2}^{1/2} = 0 \qquad (23)$$

The solution of this equation will give the pressure dependence of the hole concentration over the whole p-type region of this model. In connection with this method of characterization of a defect structure from the pressure dependence of the defect concentration, several words of warning are worthwhile. Over a restricted range where a single type of defect predominates a simple pressure dependence such as

$$[V_M'] = p = (K_{V_M'})^{1/2} P_{O_2}^{1/2}$$

or

$$[V_M''] = \frac{p}{2} (K_{V_M''}/4)^{1/3} P_{O_2}^{1/6}$$

might be observed. However, in practice it is found that data obtained on real materials are rarely free of transitions from one charge-carrier type to another or from one range of predominant defect to another. Since such transitions appear [29] to obey a simple pressure dependence over many orders of magnitude of oxygen pressure, it is easy to misinterpret a transition region and to attribute it to a simple defect structure. This point is discussed further by Bransky and Tallan [30] for the case of ThO_2. The coefficients of the model, $K_{V_M'}$, $K_{V_M''}$, and $[F]$ can be evaluated from a fit of isothermal values of p and P_{O_2} to Eq. (23). Since it is difficult to measure p directly, the electrical conductance G is measured instead, and in terms of G Eq. (23) becomes

$$G^3 \mp \frac{[F]}{\alpha} G^2 - \frac{K_{V_M'}}{\alpha^2} G P_{O_2}^{1/2} - \frac{2K_{V_M''}}{\alpha^3} P_{O_2}^{1/2} = 0 \qquad (24)$$

where α is the ratio of p to G and therefore contains the mobility and the specimen geometry. A detailed analysis of CoO using Eq. (24) is given by Bransky and Wimmer [31].

C. DOPING

Impurity atoms which are added to a crystal may occupy either the normal cation or anion lattice sites or interstitial sites, depending on the energies involved in the incorporation. Whether foreign atoms are incorporated at interstitial sites is determined primarily by size considerations. These interstitials are expected to behave like interstitial native atoms — as donors if electropositive and as acceptors if electronegative, although electronegative atoms are evidently too large to go into interstitial positions in most crystals. When foreign atoms are incorporated substitutionally their behavior is governed by the number of valence electrons compared with that of the substituted atom. If the valence is greater than the substituted atom, the impurity will behave as a donor; if the valence is smaller, the impurity will behave as an acceptor. For example, lithium in ZnO behaves as a donor when it is in interstitial positions and as an acceptor when it substitutes for Zn.

In the substitutional case the charge imbalance caused by an impurity with a valence different from that of the substituted atom can be compensated for by the formation of either ionic or electronic defects, and the type of charge compensation observed is in general dependent on the equilibrium partial pressure of oxygen. Calcium added to ZrO_2 forms an equal quantity of oxygen vacancies and forms the basis of fabricating ionic conducting ceramics. The classic example of the formation of electronic defects is Li^+ in $Ni^{2+}O^{2-}$ where a Ni^{3+} ion is formed for each Li^+ on a nickel site. At elevated temperatures these Li^+-Ni^{3+} centers are ionized, giving rise to free holes. This mechanism can therefore be used to create a well-defined concentration of charge carriers, and is termed controlled valence [28]. Figure 2 shows the exhaustion region which is observed on heating a Li-doped NiO specimen. Within this exhaustion range the hole concentration can be equated to the lithium-impurity concentration.

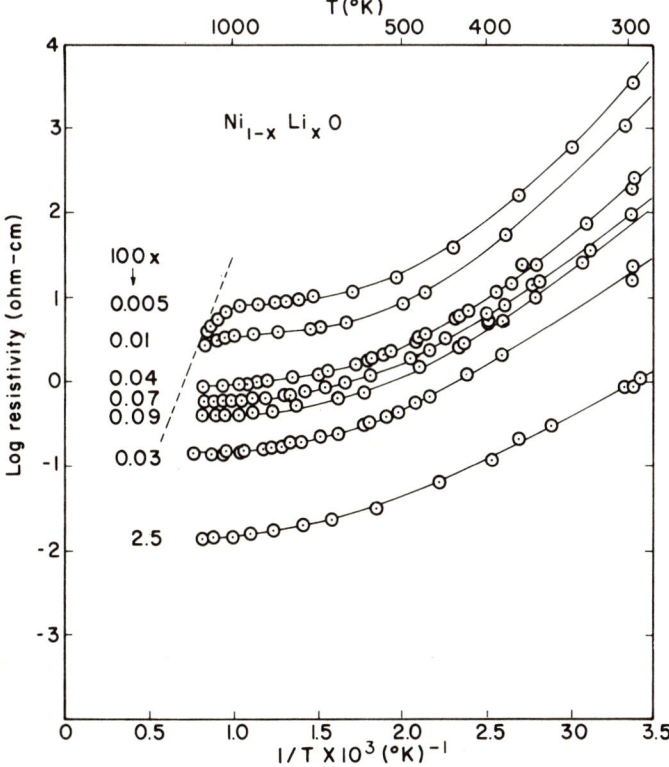

Fig. 2. The temperature dependence of the resistivity of Li-doped NiO showing an exhaustion region (after Bosman and Crevecoeur [23]).

D. ACTIVATION ENERGY

Unlike the case of ordinary doping, where the concentration of dopant is constant, the concentration of metal vacancy acceptors changes as a function of temperature and oxygen pressure. The activation energy for the electrical conductivity will therefore contain not only ionization energies, as found in the first case, but also defect formation energies. To examine the various energies contributing to the activation energy in this case, we will derive the expression for the concentration of defects from a statistical-thermodynamic point of view.

The free energy of an MO crystal containing only noninteracting V_M' and holes can be written in the form [32]

$$G = N\mu_{MO} + N_v g_v + N_v I_h - kT \ln \frac{N!}{(N_v!)^2 (N-2N_v)!} \qquad (25)$$

where N is the total number of M sites in the lattice, N_v is the number of metal vacancies, μ_{MO} is the free energy to form a MO formula unit, g_v is the free energy of neutral metal-vacancy formation (the energy required to take a cation from the interior of the crystal and place it on the surface), and I_h is the ionization energy of the metal vacancy (the energy required to remove one positive hole). The activity a_O of oxygen in the oxide can be calculated from Eq. (25) using the thermodynamic relation

$$\mu_O = \frac{\partial G}{\partial N_O} = kT \ln a_O \qquad (26)$$

and the relation between the sites

$$N = N_O = N_v + N_M \qquad (27)$$

where μ_O is the chemical potential of oxygen in the oxide and N_O and N_M are the number of oxygen and metal ions, respectively, in the crystal. In the type of experiment under consideration (oxidation or reduction of MO through an exchange of oxygen with the surrounding atmosphere) N_M remains constant. Since the departure from stoichiometry, x in MO_{1+x}, is given by the relation

$$1 + x = \frac{N_O}{N_M} \tag{28}$$

the oxygen activity in MO_{1+x} can be written as:

$$a_O = \frac{x^2}{1 - x^2} \exp\left(\frac{\mu_{MO} + g_v + I_h}{kT}\right) \tag{29}$$

When the oxide is in equilibrium with the surrounding oxygen pressure, then $a_O = P_{O_2}^{1/2}$, and from Eq. (29) we have

$$\frac{x^2}{1 - x^2} = K_{V_M'} P_{O_2}^{1/2} \tag{30}$$

where

$$K_{V_M'} = \exp\left(-\frac{\mu_{MO} + g_v + I_h}{kT}\right) \tag{31}$$

Therefore, in addition to the ionization energy I_h, the activation energy for the departure from stoichiometry also contains the formation energy of a metal vacancy and the formation energy of a formula unit of the oxide. These last two energies have opposite signs, and therefore the observed activation energy can be positive (e.g., in CoO), or negative (e.g., in MnO).

IV. EXPERIMENTAL DETERMINATION OF THE CHARGE-CARRIER CONCENTRATION AND DRIFT MOBILITY

We will now examine several experimental methods which can be used to obtain the separate parameters involved in charge-carrier transport, e.g., the magnitude and temperature dependence of the charge-carrier mobility, the free charge-carrier concentration, the activation energies, etc. Four types of measurements are commonly used to derive this information: electrical-conductivity, Seebeck-coefficient, thermogravimetric, and Hall-effect measurements. Of these we will concentrate on the first three, since the measurement of the Hall effect in most ceramics is experimentally difficult and the interpretation of the results in low-mobility materials is not as straightforward as it is in the case of broad-band semiconductors. The measurement of the conductivity of electronic conductors gives the concentration-mobility product. In order to extract the mobility, an independent measure of the charge concentration under identical conditions is required. The carrier concentration can be fixed by a known amount of dopant, the departure from stoichiometry can be measured and an electroneutrality condition assumed, or the carrier concentration can be obtained directly from the thermoelectric-power measurements. Each will be discussed separately.

A. DEPARTURE FROM STOICHIOMETRY

The departure from stoichiometry is usually obtained from weight-change measurements either under equilibrium conditions or after an isopiestic treatment [33]. Other methods, e.g., coulometric titration [34], can also be used to fix the departure. As an example, consider equilibrium thermogravimetric measurements in oxygen-excess (MO_{1+x}) oxides. If W is the balance reading at stoichiometry and W_o the true weight of the stoichiometric specimen, then

$$x = \frac{M}{16} \frac{W - W_s}{W_o} \tag{32}$$

where M is the molecular weight of the oxide. If only one type of atomic defect predominates, say V_M', then the charge-neutrality condition in Eq. (22) reduces to $p = [V_M']$, and for small departures from stoichiometry $[V_M'] = x$. [For large departures the relationship depends on how $[V_M']$ is defined, since $N_{V_M'}/N_M = x$ and $N_{V_M'}/N_O = x/(1 + x)$]. Therefore the hole mobility is given by

$$\mu_p = \frac{\sigma}{ep(A_o\rho/M)} = \frac{\sigma}{ex(A_o\rho/M)} \tag{33}$$

where A_o is Avogadro's number, ρ is the density of the crystal, and

M is the molecular weight of the crystal. Modern microbalances used for thermogravimetry can accurately measure weight changes of about ± 5 μg on a 1-g specimen. As a result, the lower limit of x that can be measured is about 50 ppm in the iron-group monoxides. The combined gravimetric-conductivity method is therefore limited experimentally to oxides with relatively large departures from stoichiometry. It should be stressed that the application of this method to obtain charge-carrier concentrations depends on an exact knowledge of the relation between the defect structure and the charge-carrier concentration. In the present example, we assumed specifically a single state of ionization with no influence of doubly ionized vacancies, impurities, etc. If, for example, we consider a region of mixed ionization where $x = [V_M'] + [V_M'']$ and $p = [V_M'] + 2[V_M'']$, then the mobility is given by

$$\mu_p = \frac{(z+1)\sigma}{(z+2)(A_o\rho/M)ex} \qquad (34)$$

where $z = [V_M']/[V_M'']$. However, the value of z is not easily obtained from the data. Such a mixed region can also result in an erroneous temperature dependence for the mobility. For the case of $p = [V_M']$ the temperature dependence of the mobility is given by

$$\frac{d \log \mu_p}{d(kT)^{-1}} = \frac{d}{d(kT)^{-1}} (d \log \sigma - d \log p)$$

$$= \frac{d}{d(kT)^{-1}} (d \log \sigma - d \log x) \qquad (35)$$

However, if we are in a mixed-ionization region the temperature dependence of x and p cannot be equated, since

$$\frac{d}{d(kT)^{-1}} (\log x - \log p)_{P_{O_2}}$$

$$= f(z)[3(\frac{d \log K_{V_M'}}{d(kT)^{-1}}) - 2(\frac{d \log K_{V_M''}}{d(kT)^{-1}})] \qquad (36)$$

where $f(z) = \frac{z}{[(6+2z)(z+1)]}$.

The maximum difference in Eq. (36) occurs for z = 3 and is given by

$$\left[\frac{d}{d(kT)^{-1}} (\log x - \log p)_{P_{O_2}}\right]_{max} \simeq \frac{1}{15}(\mu_{MO} + g_v + I_h' - 2I_h'') \quad (37)$$

This difference depends on the formation and ionization energies of the defects; however, values of 0.05-0.1 eV might easily be expected. A misinterpretation of the data in a mixed region could therefore lead to an incorrect conclusion that the mobility is thermally activated and that the charge carrier is a small polaron in the hopping regime. The conditions under which the charge-carrier mobility can be determined from combined conductivity and thermogravimetric measurements for the case $x = [V_M'] + [V_M^X]$ in the monoxides has been discussed by Bruck and Tannhauser [35].

B. SEEBECK COEFFICIENT

Another property which can be used to obtain information about the free charge-carrier concentration in a metal oxide is the thermoelectric power or Seebeck coefficient. By being directly dependent on the free charge-carrier concentration it has in fact a distinct advantage over gravimetric measurements, which are useful for the evaluation of mobilities only if the relationship between the concentration of charge carriers and point defects is known and the defect concentration is relatively large. The Seebeck coefficient is defined as

$$Q = \frac{dV}{dT} \quad (38)$$

when measured under experimental conditions such that no current flows through the specimen. If Q is taken as

$$Q = -\frac{V_h - V_c}{T_h - T_c} \quad (39)$$

where $V_h - V_c$ and $T_h - T_c$ are the emf and temperature differences between the hot and cold ends of the specimen, then the sign of the charge carrier corresponds to the sign of Q. In the case of two-carrier conduction the Seebeck coefficient is given by the relation

$$Q = \frac{\sigma_n Q_n + \sigma_p Q_p}{\sigma} \qquad (40)$$

If, however, the electrical conductivity can be ascribed to a single type of charge carrier, the Seebeck coefficient can be written [36] as

$$Q = \pm \frac{k}{e} \left(\frac{E_F}{kT} + A\right) \qquad (41)$$

where E_F is the Fermi energy measured relative to the transport level and A is an energy-transport term. For a nondegenerate semiconductor where Boltzmann statistics can be applied Eq. (41) has the form

$$Q = \pm \frac{k}{e} \left(\ln \frac{p_o}{c} + A\right) \qquad (42)$$

where c is the charge carrier concentration and p_o the effective density of states at the transport level. Thus the determination of the free charge-carrier concentration from the thermal-emf measurements requires a knowledge of both the effective density of states p_o and the transport-energy term A. In addition, the determination is rather insensitive because of the logarithmic relation between Q and c. These disadvantages are partially compensated by the fact that it is generally easier to make accurate Seebeck-coefficient measurements than gravimetric ones.

In order to determine values for p_o and A, two extreme cases can be distinguished. In Case I the charge carriers move in a wide spherical band giving

$$p_o = 2\left(\frac{2\pi m^* kT}{h^2}\right)^{3/2} = 4.835 \times 10^{15} \left(\frac{m^*}{m} T\right)^{3/2} \text{ cm}^{-3}$$

and AkT is the kinetic energy of the charge carrier. If it is assumed that the mean free path is independent of energy, then $A = 2$. In Case II the band is very narrow and all the available states are included in an interval kT. Now the density of states is one per available site for the charge carrier, so that if all the cations are equivalent available sites p_o is of the order of 10^{22} cm^{-3} and exceeds the density of states in a broad-band semiconductor by several orders of magnitude. Furthermore, for a very narrow polaron band the kinetic energy of the charge carrier is much smaller than kT and $A \simeq 0$ [37]. For a hopping model, Heikes, Maradudin, and Miller [38] and Austin and Mott [17] have shown that $A = (1-2\beta)W_h/kT$, where $0 < \beta < \frac{1}{2}$ is a constant. The condition $\beta \neq \frac{1}{2}$ arises from the assumption that the energy of the charge carrier on the initial site is different from its energy on the adjacent site into which it jumps. If instead the hopping energy W_h is equally divided between the sites, then $\beta = \frac{1}{2}$ and $A = 0$. In a study of hopping charge carriers in Eu_3S_4, Bransky, Tallan, and Hed [39] found $W_h = 0.16$ eV, $\beta = 0.4$, and $dA/d(kT)^{-1} = 0.03$ eV.

Combining Eqs. (1) and (42) we obtain

$$\frac{d \ln \sigma}{d(kT)^{-1}} \pm \frac{e}{k} \frac{dQ}{d(kT)^{-1}} = \frac{d}{d(kT)^{-1}} (\ln \mu + \ln p_o + A) \qquad (43)$$

Therefore, when the temperature dependence of p_o and A can be neglected (or are known) the temperature dependence of μ can be obtained from this equation. Such an analysis was used to demonstrate the absence of an activated-hole mobility in Li-doped NiO [23] and CoO [24] and nonstoichiometric CoO [31] and NiO [25]. However, Eq. (43) is somewhat crude and should not be used to obtain activation energies to better than about kT, since in the case of, e.g., a broad band, $d \ln p_o/d(kT)^{-1} = -(3/2)kT$, which averages to 0.15 eV in the temperature range 1000-1600°K, a range of considerable interest in ceramic research.

V. ANALYSIS OF ELECTRICAL-CONDUCTIVITY AND THERMAL emf MEASUREMENTS FOR TWO TYPES OF CHARGE CARRIERS

The analysis of the general case, where both electrons and holes contribute to the electrical conduction and thermal emf, will be discussed. It is assumed that only electronic conduction takes place (this is not generally true, especially at high temperatures where careful elimination of the ionic contribution near stoichiometry must be made in order for the following analysis to be valid). It is also assumed that the states of the semiconductor are non-degenerate ($\Delta E > 3kT$ is generally accepted as a good criterion for this assumption) and, furthermore, that μ_p, μ_n, A_p, and A_n depend on temperature only. Equations (2) and (40) give the total conductivity and Seebeck coefficient, respectively. In Eq. (40)

$$Q_n = -\frac{k}{e}(\ln \frac{N_c}{n} + A_n)$$

$$Q_p = +\frac{k}{e}(\ln \frac{N_v}{p} + A_p) \qquad (44)$$

where N_c and N_v are the effective densities of states at the conduction- and valence-band levels, respectively. Becker and Frederikse [40] showed that σ/σ_i, Q, and R (the Hall constant) can be expressed in terms of the variable $\alpha = p/bn = \sigma_p/\sigma_n$ where $b = \mu_n/\mu_p$ and the intrinsic conductivity is $\sigma_i = n_i e(\mu_n + \mu_p)$. Therefore

$$\frac{\sigma}{\sigma_i} = \frac{b^{1/2}(1 + \alpha)}{\alpha^{1/2}(1 + b)} \qquad (45a)$$

and

$$(\frac{2e}{k})Q = \frac{\epsilon(\alpha - 1)}{\alpha + 1} + \delta - \ln \alpha \qquad (45b)$$

where

$$\epsilon = \frac{E_g}{kT} + A_p + A_n$$

$$\delta = A_p - A_n + \ln \frac{N_v}{bN_c}$$

Figures 3 and 4 show plots of log (σ/σ_i) and of Q as functions of log α at constant temperature. At $\alpha = 1$, i.e., $\sigma_n = \sigma_p$, a minimum of σ/σ_i is observed and at this point:

$$\frac{\sigma_{min.}}{\sigma_i} = \frac{2b^{1/2}}{b+1} = \frac{2}{(\mu_p/\mu_n)^{1/2} + (\mu_n/\mu_p)^{1/2}} \tag{46}$$

Since the mobility ratio is assumed to be independent of temperature, σ_i is proportional to σ_{min} and the gap energy E_g can be obtained from the slope of a plot of log σ_{min} vs $(kT)^{-1}$. Since E_g is temperature dependent to a first approximation as $E_g = E_o(0°K) - \beta T$, only E_o is obtained from this analysis. Becker and Frederikse [40] have used the above results to analyze semiconductors in which the concentration of charge carriers at high temperature due to non-stoichiometry can be expressed as

$$n = C_n(P)^{-1/y} \exp\left(-\frac{W_n}{kT}\right) \tag{47}$$

where P is the equilibrium partial pressure of one of the constituents. Since $np = n_i^2$ we also have

$$p = C_p(P)^{1/y} \exp\left(-\frac{W_p}{kT}\right) \tag{48}$$

where $W_p + W_n = E_g$ and $C_n C_p = N_v N_c$. The variable $\alpha = p/bn$ now becomes

$$\alpha = C_n^{-2}\left(\frac{N_v N_c}{b}\right)(P)^{2/y} \exp\left(-\frac{E_g - 2W_n}{kT}\right) \tag{49}$$

At constant temperature and y, log α = const. + (2/y) log P, so the same analysis described above for plots of Q and log (σ/σ_i) vs log α can be applied to Q and log (σ/σ_i) vs log P plots. From such an analysis it is found that

ELECTRONIC CONDUCTION 295

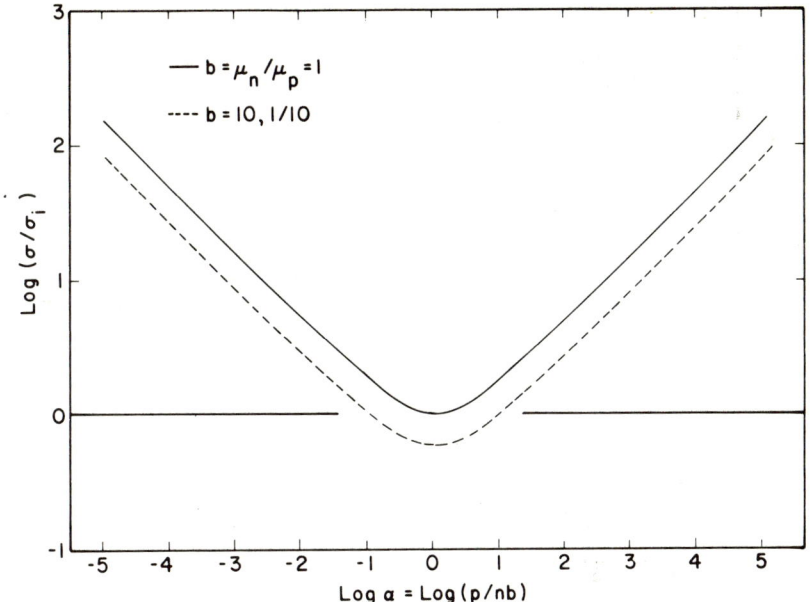

FIG. 3. Log σ/σ_i vs log α at constant temperature (after Becker and Frederikse[40].

$$\log \frac{P(\text{at } Q_{max})}{P(\text{at } Q_{min.})} = y \log 2(\epsilon - 1) \quad (50)$$

which indicates that a rather large pressure range is required to cover both the maximum and minimum of Q for compounds with a large energy gap. For example, in TiO_2, $E_g \simeq 3$ eV, $\epsilon \simeq 50$ and $y = 6$ at 800°K so that $P(\text{at } Q_{max})/P(\text{at } Q_{min.}) = 10^{12}$. The above analysis has been applied to doped TiO_2 [40,41] (see Fig. 5) and to pure MnO [42].

A somewhat different approach to the two-carrier analysis was used by Jonker [43], who showed that for nondegenerate semiconductors with the simplified assumptions made above the relation between Q and σ is given by

$$Q = \pm \frac{k}{2e} \epsilon (1 - \frac{\sigma_{min.}^2}{\sigma^2})^{1/2} - \frac{k}{e} \ln \{\frac{\sigma}{\sigma_{min.}} [1 \pm (1 - \frac{\sigma_{min.}^2}{\sigma^2})^{1/2}]\} + \frac{k}{e} \ln \delta \quad (51)$$

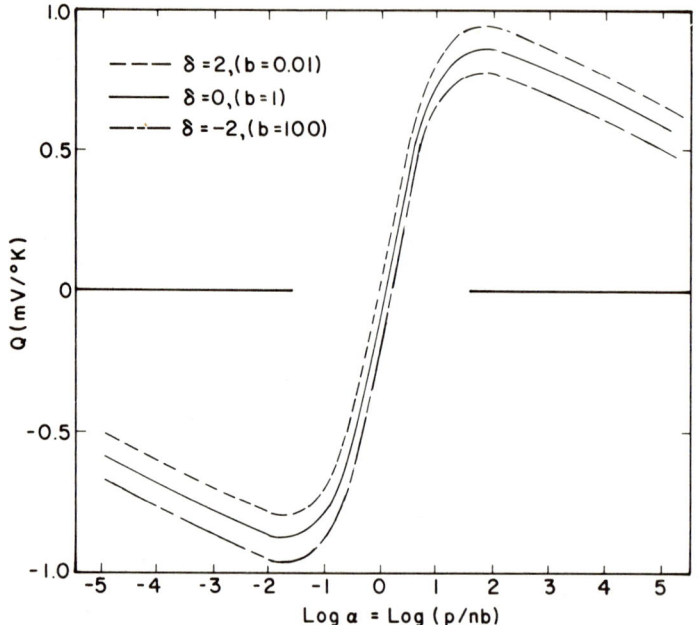

FIG. 4. Log Q vs log α at constant temperature (after Becker and Frederikse [40]).

where ϵ, δ, and σ are the same parameters defined in Eq. (45). When Q is plotted as a function of log (σ/σ_{min}) the resulting curve has the shape of a pear and is referred to as "Jonker's pear". Jonker's pear curves are shown in Fig. 6 for $La_{0.9}Y_{0.1}CoO_3$ [20] and in Fig. 7 for MnO [44].

An important feature of the pear is that it depends only on the intrinsic parameter. Thus, the correctness of a simultaneous measurement of σ and Q can be checked by their fit to the appropriate Jonker pear. This fit is independent of doping, departure from stoichiometry, and other details of the defect structure.

VI. EXAMPLES IN CERAMIC MATERIALS

Examples of the charge-transport properties of some specific ceramic materials will now be discussed. For this purpose we will

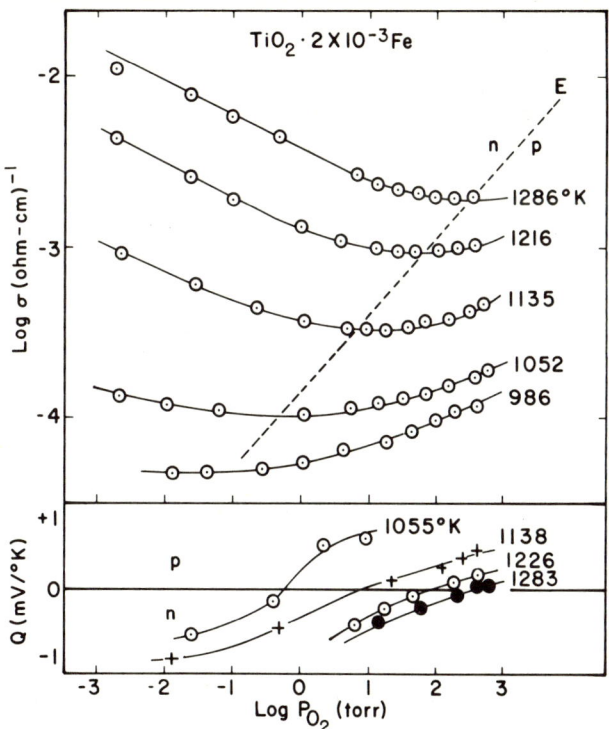

FIG. 5. Electrical conductivity and Seebeck coefficient of Fe-doped TiO_2 as a function of oxygen pressure (data of Ref. 41 analyzed by Becker and Frederikse [40]).

arbitrarily group together materials similar enough in origin or behavior to simplify the discussion. The experimental work that has been performed on these materials has certainly resulted in a better understanding of the basic theory of point defects and charge transport, since many different effects have been observed and adequately explained. The problem has been not only that it is often extremely difficult to reproduce exactly the results of a previous investigation, but also that the experiments themselves are often not sensitive enough to unquestionably differentiate between alternative explanations. As a result, the general behavior of most materials has certainly been characterized, but the specific details of the defect structure and charge-transport mechanisms are still being argued.

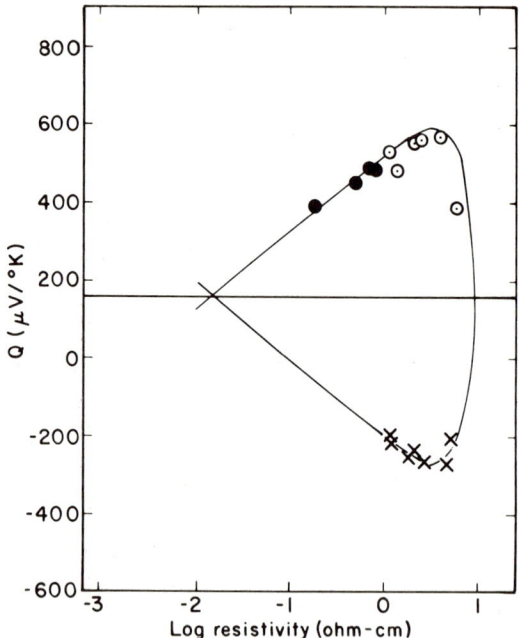

FIG. 6. Jonker's pear curve for $La_{0.9}Y_{0.1}CoO_3$. X — Mn doped; ● — Sr doped; ⊙ — undoped (after Jonker [20]).

A. Al_2O_3, MgO, BeO

This group consists of the ceramics which are relatively good insulators even at high temperature. This low conductivity results at least in part from a small concentration of charge carriers due to a large band gap (a normal ionic compound, see Section III) and small departure from stoichiometry. Because of this small departure from stoichiometry there are no observable weight changes and the high resistivity makes conductivity and Seebeck-coefficient measurements experimentally difficult. Because of the low electronic conductivity, ionic-conduction and impurity effects often become the dominating influences although the possibility of intentional doping to fix the carrier concentration at a known level has not been extensively investigated. All of the studies point to the need for good measurements on high-purity material. However, at high temperature impurity pick-up in situ is usually unavoiable, so there is no

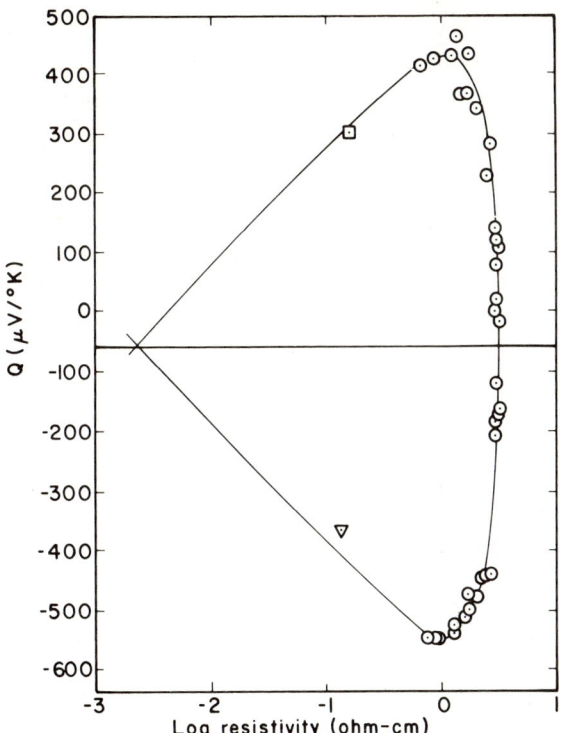

FIG. 7. Jonker's pear curve for MnO (after Crevecoeur and de Wit [44]).

assurance that the availability of high-purity specimens will actually lead to experimental measurements free from impurity effects.

1. Al_2O_3

Because of its great technological importance alumina has received considerable attention, but there is still controversy as to whether the true high-temperature electronic properties have ever been measured. Although electrical-conductivity measurements up to 1700°C [45] were originally interpreted in terms of electronic conduction, Peters et al. [46] found that the conductance measured with or without a specimen in the holder was about the same. This observation has been attributed to "gas-phase conduction," and attempts have been made to remove this extraneous current path by hermetically sealing off the gas on one side of the specimen from that on the other. However, the phenomenon is not well understood and needs further investigation. There is considerable evidence from galvanic-cell measurements [47,48] below 1200°C and from a correspondence between activation energies observed in electrical measurements and self-diffusion experiments [49,50] that ionic conduction is a major contributor to the total conductivity at least in the temperature range below 1200-1300°C. Earlier galvanic-cell measurements indicated that conduction is predominantly electronic above 1500°C. However, Brook et al. [51] recently discussed the possible experimental errors in high-temperature electrical measurements and proposed that ionic conduction may predominate to much higher temperatures. This view is substantiated by recent galvanic-cell measurements by Yee and Kroger [83], who guarded against gas-phase and surface-leakage currents and observed that ionic conduction predominates to over 1500°C in single crystals but that polycrystals have a large electronic contribution.

2. MgO

There is also some controversy about the electronic properties of this material. Mitoff [52] used MgO as an electrolyte in a galvanic cell, guarded against surface and gas-phase conduction, and found that the conduction is primarily ionic below 1000°C, with a transition to electronic conduction above 1500°C. The source of the electronic carriers is not clear but is evidently due to variable-valence impurities [52,53]. In contrast, on the basis of the similarity of activation energies between electrical-conductivity and self-diffusion measurements, Davies [54] proposed that only ionic conduction prevails. Although such an argument can certainly be based on activation energies, it is not direct evidence, and the careful experimental work of Mitoff cannot be discounted. Osburn and Vest [55] have measured the oxygen partial-pressure dependence of the electrical conductivity up to 1600°C and interpret their measurements in terms of electronic charge carriers arising from a departure from stoichiometry. As in the case of alumina, it is difficult to speculate on the electronic transport mechanism.

3. BeO

Beryllium oxide has not been studied as extensively as either alumina or magnesia. Cline et al. [56] used a galvanic cell guarded against gas-phase conduction and found BeO to be a 100%-ionic conductor up to 1700°C at high oxygen pressure (10^{-3} atm). The concentration of the ionic defects responsible for the conduction is evidently controlled by the impurity content [57], especially aluminum and silicon.

B. IRON-GROUP MONOXIDES

The studies of electronic transport in these oxides, which have been extensively reviewed by Bosman and van Daal [11] and Adler [58], offer good examples of the various experimental approaches which can be used. A major portion of the research has been directed toward obtaining the magnitude and temperature dependence of the drift mobility, since this is one of the few critical tests which have been proposed to determine whether the conduction is band-type or small-polaron. The monoxides of cobalt, nickel, and iron exhibit predominantly p-type conduction, while manganous oxide becomes n-type at low equilibrium oxygen pressure because of the presence of highly mobile electrons. The maximum departure from stoichiometry occurs at high oxygen pressure and in CoO, MnO, and FeO is easily measured by gravimetric techniques. Although considerably smaller, the departure in NiO is also measurable by this technique. The observed weight change has been used to estimate the electron-hole concentration and, combined with the conductivity, the drift mobility. Lithium doping has also been used to fix the electron-hole concentration at a known value. As the temperature is increased these acceptors are thermally ionized and an exhaustion region occurs.

Because of the small departure from stoichiometry in NiO the exhaustion region is evident in a resistivity vs 1/T curve, while in the other oxides the region is masked by other sources of charge carriers, such as native nonstoichiometric defects. The exhaustion region in NiO is shown in Fig. 2, where from about 400 to 725°C the resistivity becomes nearly independent of temperature. The finite slope in this range amounts to 0.1-0.15 eV, indicating that if NiO were a small-polaron conductor the hopping energy could be no larger than this value. In the exhaustion region $p \simeq [Li']$, and the drift mobility can be determined by combining the concentration of dopant with the experimental value of resistivity. The resistivity has been observed to vary linearly with dope in the exhaustion region for MnO, NiO, CoO, and α-Fe_2O_3, and the values of drift mobility at 1200°K are 0.02, 0.4, 0.25, and 0.1 cm^2/V-sec, respectively [11]. The values obtained from combined electrical-conductivity and thermogravimetric measurements are 0.15 cm^2/V-sec for MnO [42] at 1200°C, 0.25 cm^2/V-sec for NiO [59] at 1400°C, and 0.4 cm^2/V-sec for CoO [60] at 1350°C. Therefore, with the exception of MnO, there is good agreement between the experiments where the electron-hole concentration is fixed by a known quantity of lithium and those where the defect concentration is determined by thermogravimetric techniques and an electroneutrality condition assumed. Attempts to measure the temperature dependence of the mobility using these combined-property measurements has resulted in somewhat conflicting results. At high temperature the magnitude of kT is relatively large, and since the experimental error in a measured temperature dependence is probably not less than this amount it is difficult to determine the difference between two temperature dependencies to the accuracy required to measure an activation energy for mobility of less than about 0.2 eV. The high-temperature measurements are necessary, however, to study the nonstoichiometric defect structure under equilibrium conditions at these temperatures.

The electrical conductivity of MnO exhibits a definite minimum as it changes from p-type at high oxygen pressure to n-type at low oxygen pressure. The Seebeck coefficient also passes through zero and the data are amenable to a Becker and Frederikse [40] or Jonker [43] two-carrier analysis. From such an analysis Hed and Tannhauser [42] found that the ratio of electron to hole drift mobilities was at least 10. Hall-mobility measurements obtained on pure [61] and lithium-doped [62,63] MnO indicate a large Hall mobility for the electrons, on the order of 10 cm^2/V-sec. However, there is disagreement on the value of the Hall mobility for holes. Measurements on the Li-doped specimens indicate a value of about 0.007 cm^2/V-sec, whereas Hall measurements made on pure material as a function of the departure from stoichiometry require a two-carrier analysis and result in a value of hole mobility of 0.2 cm^2/V-sec. Crevecoeur and de Wit [44] also measured the Seebeck coefficient and electrical conductivity of MnO and obtained a fit to Jonker's pear at 1100°C. Using the Hall-mobility ratio obtained on Li-doped specimens, i.e., $\mu_n/\mu_p \simeq 1000$, they obtain 1.8 eV as an upper limit of the band-gap energy at 1100°C. The analysis of the conductivity minimum according to Eq. (46) resulted in a gap energy of 2.4 eV. As pointed out in Section V, the use of Eq. (46) results in the gap energy at 0°K and, therefore, assuming a linear dependence on temperature, the energy gap can be written as $E_g = 2.4(1-2 \times 10^{-4}T)$ eV.

Along with UO_{2+x}, FeO_{1+x} is one of the few oxides on which direct evidence of the defect structure has been obtained by neutron-diffraction studies. This evidence [64] indicates that the predominant defect is a complex consisting of an iron vacancy pair associated with an iron interstitial. This complex defect is evidently the direct result of the unusually large amount of oxygen excess found in FeO, ranging from a minimum of 4.5% up to 15%. Combining the conductivity [65,66] and thermogravimetric [67] data, a hole mobility on the order of 0.14 cm^2/V-sec is obtained [68]. This calculation assumes that all the iron vacancies are doubly ionized, which is not the correct defect model, although it should still give the right order of magnitude.

C. ZrO_2, HfO_2, ThO_2

These oxides are similar in that at high oxygen pressure they are predominantly p-type electronic conductors, but as the oxygen pressure is decreased they exhibit a considerable amount of ionic conduction. At very low oxygen pressure, combined with high temperatures, the conduction then becomes n-type electronic. The ionic range can be enhanced by doping with lower-valent cations; this forms the basis for obtaining pure ionic conductors for use as electrolytes over a wide range of oxygen partial pressure (see Chapter 7). However, a region of ionic behavior is observed even in the purest materials available, so it is of interest to discuss possible explanations for this behavior.

Gravimetric measurements [69] indicate a small but measurable weight change in HfO_2 at high oxygen pressure. In ZrO_2 and ThO_2 no weight changes are observed at high oxygen pressure [70] but in ZrO_2 a measurable change has been reported [71] to occur close to the $Zr-ZrO_2$ phase boundary. Corresponding measurements for departure from stoichiometry at the $Th-ThO_2$ and $Hf-HfO_2$ phase boundaries have not been reported. The observation that a weight change is not measurable at high oxygen pressure even with a sensitive microbalance sets an upper limit of about $2-3 \times 10^{18}$ cm^{-3} for the concentration of excess oxygen. Assuming that the nonstoichiometric defects are fully ionized this limit can be used to estimate a corresponding electron-hole concentration, and combining this with the measured electronic conductivity a lower limit on the hole drift mobility can be set. This lower limit is 1.5×10^{-4} cm^2/V-sec for ZrO_2 at 1000°C and 1 atm (i.e., monoclinic ZrO_2) and about 1.5×10^{-3} cm^2/V-sec for ThO_2. In HfO_2, Tallan et al. [69] combined thermogravimetric and conductivity measurements and calculated a hole mobility which varied from 1.7×10^{-3} cm^2/V-sec at 1500°C to 3.8×10^{-4} cm^2/V-sec at 1000°C.

The Seebeck coefficient of all three of these materials has been measured, although the analysis is complicated [72] by the presence of ionic conduction. Close to 1 atm oxygen pressure the conduction is predominantly p-type electronic, so only the Seebeck coefficient in this region will be discussed. At 1000°C Tallan and Bransky [72] find that, in ThO_2 at 10^{-1} atm, Q_p = 1.4 mV/°K. Assuming that A = 0 and p_o = [Th] = 2.2 x 10^{22} cm^{-3}, a hole concentration of 1.9 x 10^{15} cm^{-3} is obtained using Eq. (41). Combining this value with a hole conductivity of 1.6 x 10^{-3} ohm^{-1} cm^{-1}, a hole drift mobility of 5 cm^2/V-sec is obtained. At 1200°C a similar calculation results in a value of 2 cm^2/V-sec. In ZrO_2 [73] at 1000°C and 1 atm, Q_p is also about 1.4 mV/°K. Since the conductivity of ZrO_2 under these conditions is about 10^{-4} ohm^{-1} cm^{-1}, the calculated hole drift mobility is about 0.8 cm^2/V-sec. In HfO_2 [73] at 1000°C and 1 atm, Q_p = 0.8 mV/°K and σ_p = 2.8 x 10^{-4} ohm^{-1} cm^{-1}, which results in a mobility of 9.8 x 10^{-4} cm^2/V-sec. At 1200°C, the hole drift mobility is 1.4 x 10^{-3} cm^2/V-sec. Although these small values of hole mobility for HfO_2 agree very well with the combined thermogravimetric-conductivity measurements, they should be used with caution, since neither method is a direct measurement of mobility.

In all three of these oxides a strong dependence of the conductivity on oxygen pressure is observed on both sides of the ionic region. These pressure dependencies have usually been interpreted in terms of departures from stoichiometry, i.e., on the p-type side $p = z \ [V_M^{z'}]$ or $p = z \ [O_i^{z'}]$ and on the n-type side $n = z \ [V_O^{z\cdot}]$ or $n = z \ [M_i^{z\cdot}]$. If this type of neutrality condition is correct, then the electron or hole concentration is on the order of the ionic defect concentration, and therefore if there is a transition to a mixed ionic and electronic conduction region this must indicate that both the ions and electronic carriers have the same order of magnitude of mobility. On this basis the observation of small hole mobilities is reasonable. One alternative to the above interpretation of the oxygen-pressure dependence is an impurity-controlled model. In this case the effective charge state of the impurities enters into the neutrality condition and the electronic carrier concentration may be orders of magnitude lower than the ionic defect concentration. This type of model has been used [30] to explain the pressure dependence observed in ThO_2 close to 1 atm oxygen pressure.

D. TiO_{2-x}, Nb_2O_{5-x}, CeO_{2-x}, AND UO_{2+x}

These oxides all exhibit a rather large departure from stoichiometry, and, except for UO_2, are n-type semiconductors and exhibit transitions to lower oxides at low oxygen pressure. In TiO_{2-x} and Nb_2O_{5-x} the formation of shear structures or Magnelli phases has been suggested at large departures from stoichiometry. A considerable amount of research has been done on these oxides, with a great portion devoted to attempts to identify the predominant defect structure.

Rutile (TiO_2) has received considerable attention since it has been available in single-crystal form for many years. Blumenthal et al. [74] combined recent conductivity and gravimetric measurements to show that the electron drift mobility is about 0.1 cm^2/V-sec and essentially independent of temperature over the range 1350-1500°K. Hall-mobility measurements by Bransky and Tannhauser [75] are in good agreement with this drift-mobility value in the overlapping temperature range. The agreement between the magnitude of the Hall and drift mobilities and the fact that the Hall mobility decreases rapidly with temperature are not consistent with a small-polaron model. Many of the charge-transport effects observed at low temperatures (below 800°C) are felt [76,77] to be due to impurities, especially hydrogen.

Janninck and Whitmore [78] studied the electrical conductivity and Seebeck coefficient of nonstoichiometric Nb_2O_5 at constant composition using an isopiestic technique. In this technique the departure from stoichiometry is fixed under equilibrium conditions at high temperature and then the properties are measured at lower temperatures, where it is assumed that the specimen retains its high-temperature composition. The magnitude of the departure from stoichiometry was measured by gravimetric techniques and combined with the conductivity to give an electron drift mobility of 0.2 cm^2/V-sec at 1000°C and a negative temperature dependence of about $T^{-0.5}$. Using equilibrium thermogravimetric and electrical-conductivity measurements, Wimmer and Tripp [79] confirmed these results. The Seebeck [78] and electrical-conductivity [33] data were interpreted as indicating that the electrons move in a narrow 4d band.

The dependence of the n-type electrical conductivity of CeO$_2$ on temperature and oxygen pressure has been studied to determine the predominant nonstoichiometric defect. From a computer fit to electrical-conductivity data, Blumenthal et al. [80] attributed 0.18 eV to the motion of small polarons. A combined study of conductivity, Seebeck coefficient and weight change has not been published.

Aronson et al. [81] studied the electrical conductivity and Seebeck coefficient of UO$_{2+x}$ at low temperatures, where specimens of constant composition could be used. The data were analyzed on the basis of a small-polaron model with a mobility of 0.055 cm^2/V-sec at 1100°C and an activation energy of 0.3 eV. The predominant defect [82] in UO$_2$ is an oxygen-vacancy/oxygen-interstitial complex, and the hole transfer occurs between U^{4+} and U^{5+} ions. It remains to be seen whether further investigations will continue to support a small-polaron model for CeO$_2$ and UO$_2$.

REFERENCES

1. J. Bardeen and W. Shockley, Phys. Rev., 80, 72 (1950); see also W. Shockley, Electrons and Holes in Semiconductors, Van Nostrand, New York, 1950, pp. 264-278, 520.

2. E. M. Conwell and V. F. Weisskopf, Phys. Rev., 77, 388 (1950).

3. L. D. Landau, Phys. Z. Sowjetunion, 3, 644 (1933).

4. J. Appel, Solid State Physics, Vol. 21 (F. Seitz, D. Turnbull, and H. Ehrenreich, eds.), Academic Press, New York, 1968, pp. 193-391.

5. D. J. Howarth and E. H. Sondheimer, Proc. Royal Soc. Ser. A,219, 53 (1953).

6. R. L. Petritz and W. W. Scanlon, Phys. Rev., 97, 1620 (1958).

7. F. Low and D. Pines, Phys. Rev., 98, 414 (1958).

8. R. P. Feynman, Phys. Rev., 97, 660 (1955).

9. R. P. Feynman, R. W. Heilwarth, C. K. Iddings, and P. M. Platzman, Phys. Rev., 127, 1004 (1962).

10. P. Gerthsen, E. Kauer, and H. G. Reik, in Festkörper — probleme, Vol. 5 (F. Sauter, ed.), Friedr. Vieweg und Sohn, Braunschweig, West Germany, 1966, pp. 1-56.

11. A. J. Bosman and H. J. van Daal, Advan. Phys., 19, 1 (1970).

12. T. Holstein, Ann. Phys., 8, 343 (1959).

13. L. Friedman, Phys. Rev., 135, 233 (1964).

14. H. Frolich, Advan. Phys., 3, 325 (1954).

15. D. M. Eagles, J. Phys. Chem. Solids, 25, 1243 (1964).

16. D. Emin and T. Holstein, Ann. Phys. (N. Y.), 53, 439 (1969).

17. I. G. Austin and N. F. Mott, Advan. Phys., 18, 41 (1969).

18. N. F. Mott, J. Non-cryst. Solids, 1, 1 (1968).

19. P. W. Anderson, Phys. Rev., 109, 1492 (1958).

20. G. H. Jonker, Philips Res. Rept., 24, 1 (1969).

21. F. J. Morin, Bell System Tech. J., 37, 1047 (1958).

22. R. R. Heikes and W. D. Johnston, J. Chem. Phys., 26, 582 (1957).

23. A. J. Bosman and C. Crevecoeur, Phys. Rev., 144, 763 (1966).

24. A. J. Bosman and C. Crevecoeur, J. Phys. Chem. Solids, 30, 1151 (1969).

25. I. Bransky and N. M. Tallan, in Physics of Electronic Ceramics (L. L. Hench and D. B. Dove, eds.), Marcel Dekker, New York, 1968, pp. 67-98.

26. F. A. Kroger, J. Phys. Chem. Solids, 29, 1889 (1968).

27. S. van Houten, J. Phys. Chem. Solids, 17, 7 (1960).

28. E. J. Verwey, P. W. Haaijman, F. C. Romeijn, and G. W. van Oosterhout, Philips Res. Rept., 5, 173 (1950).

29. C. J. Kevane, Phys. Rev., 133, A1431 (1964).

30. I. Bransky and N. M. Tallan, J. Amer. Ceram. Soc., 53, 625 (1971).

31. I. Bransky and J. M. Wimmer, J. Phys. Chem. Solids, 33, 801 (1972).

32. G. G. Libowitz, in Mass Transport in Oxides (J. B. Wachtman and A. D. Franklin, eds.), NBS Special Pub. 296, U. S. Govt. Printing Office, Washington, D.C., 1968, pp. 109-118.

33. R. F. Janninck and D. H. Whitmore, J. Chem. Phys., 37, 2750 (1962).

34. H. F. Rizzo, R. S. Gordon, and I. B. Cutler, in Mass Transport in Oxides (J. B. Wachtman and A. D. Franklin, eds.), NBS Special Pub. 296, U. S. Govt. Printing Office, Washington, D.C., 1968, pp. 129-142.

35. A. Bruck and D. S. Tannhauser, J. Appl. Phys., 37, 3647 (1966).

36. A. F. Ioffe, Physics of Semiconductors, Infosearch, London, 1960.

37. G. L. Sewell, Phys. Rev., 129, 597 (1963).

38. R. R. Heikes, A. A. Maradudin, and R. C. Miller, Ann. Phys. (Paris), 8, 733 (1963).

39. I. Bransky, N. M. Tallan, and Z. Hed, J. Appl. Phys., 41, 1787 (1970).

40. J. H. Becker and H. P. R. Frederikse, J. Appl. Phys. Suppl., 33, 447 (1962).

41. J. Rudolph, Z. Naturforsch., 14a, 727 (1959).

42. Z. Hed and D. Tannhauser, J. Chem. Phys., 47, 2090 (1967).

43. G. H. Jonker, Philips Res. Rept., 23, 131 (1968).

44. C. Crevecoeur and H. J. de Wit, Solid State Commun., 6, 295 (1968).

45. J. D. Pappis and W. D. Kingery, J. Amer. Ceram. Soc., 44, 459 (1961).

46. D. W. Peters, L. F. Feinstein, and C. Peltzer, J. Chem. Phys., 42, 2345 (1964).

47. T. Matsumura, Canad. J. Phys., 44, 1685 (1966).

48. W. J. Lackey, in *Materials Science Research*, Vol. 5 (H. Palmour III and W. W. Kriegel, eds.), Plenum Press, New York, 1971, pp. 489-502.

49. N. M. Tallan and H. C. Graham, *J. Amer. Ceram. Soc.*, 48, 512 (1965).

50. M. O. Davies, Nat. Aeronautics and Space Admin., Rept. NASA TN D-2765, 1965.

51. R. J. Brook, J. Yee, and F. A. Kroger, *J. Amer. Ceram. Soc.*, 9, 444 (1971).

52. S. P. Mitoff, *J. Chem. Phys.*, 36, 1383 (1962); 41, 2561 (1965).

53. S. P. Mitoff, *J. Chem. Phys.*, 31, 1261 (1959).

54. M. O. Davies, *J. Chem. Phys.*, 38, 2047 (1963).

55. C. M. Osburn and R. W. Vest, *J. Amer. Ceram. Soc.*, 9, 428 (1971).

56. C. Cline, J. Carlburg, and H. W. Newkirk, *J. Amer. Ceram. Soc.*, 50, 55 (1967).

57. C. F. Cline and H. W. Newkirk, *J. Chem. Phys.*, 49, 3496 (1968).

58. D. Adler, in *Solid State Physics*, Vol. 21 (F. Seitz, D. Turnbull, and H. Ehrenreich, eds.), Academic Press, New York, 1968, pp. 1-113.

59. I. Bransky and N. M. Tallan, *J. Chem. Phys.*, 49, 1243 (1968).

60. B. Fisher and D. Tannhauser, *J. Chem. Phys.*, 44, 1663 (1966).

61. M. Gvishi, N. M. Tallan, and D. S. Tannhauser, *Solid State Commun.*, 6, 135 (1968).

62. P. Nagels and M. Denayer, *Solid State Commun.*, 5, 193 (1967).

63. H. J. de Wit and C. Crevecoeur, *Phys. Letters*, 25A, 393 (1967).

64. W. L. Roth, *Acta. Cryst.*, 13, 140 (1960).

65. G. H. Geiger, R. L. Levin, and J. B. Wagner, Jr., *J. Phys. Chem. Solids*, 27, 947 (1966).

66. I. Bransky and D. S. Tannhauser, *Trans. AIME*, 239, 75 (1967).

67. I. Bransky and A. Z. Hed, *J. Amer. Ceram. Soc.*, 51, 231 (1968).

68. I. Bransky and D. S. Tannhauser, Physica, 37, 547 (1967).

69. N. M. Tallan, W. C. Tripp, and R. W. Vest, J. Amer. Ceram. Soc., 50, 279 (1967).

70. W. C. Tripp, private communication.

71. F. J. Keneshea and D. L. Douglass, Oxid. Metals, 3, 1 (1971).

72. N. M. Tallan and I. Bransky, J. Electrochem. Soc., 118, 345 (1971).

73. N. M. Tallan, private communication.

74. R. N. Blumenthal, J. C. Kirk, and W. M. Hirthe, J. Phys. Chem. Solids, 28, 1077 (1967).

75. I. Bransky and D. S. Tannhauser, Solid State Commun., 7, 245 (1969).

76. G. J. Hill, Brit. J. Appl. Phys., 1, Ser. 2, 1151 (1968).

77. L. J. Van Ruyven and J. D. Chase, Appl. Phys. Letters, 12, 214 (1968).

78. R. F. Janninck and D. H. Whitmore, J. Chem. Phys., 39, 179 (1963).

79. J. M. Wimmer and W. C. Tripp, unpublished data.

80. R. N. Blumenthal and R. J. Panlener, J. Phys. Chem. Solids, 31, 1190 (1970).

81. S. Aronson, J. E. Rulli, and B. E. Schaner, J. Chem. Phys., 35, 1382 (1961).

82. B. T. M. Willis, Proc. Brit. Ceram. Soc., 1, 9 (1964).

83. J. Yee and F. A. Kröger, J. Am. Ceram. Soc., 56, 189 (1973).